Earth, Water, Wind, and Sun

Also by D. S. Halacy, Jr.

Earth, Water, Wind, and Sun

Our Energy Alternatives

D. S. Halacy, Jr.

Harper & Row, Publishers

New York, Hagerstown, San Francisco, London

This book is for Paul

[handwritten notes:] E,W,W, and S: OEA
Halacy, Jr. D.S. E,W,W, and S: OEA. N.Y.: Harper + Row, 1977.

FIRST EDITION

Designed by C. Linda Dingler

Library of Congress Cataloging in Publication Data
Halacy, Daniel Stephen, 1919–
  Earth, water, wind, and sun, our energy alternatives.
  Bibliography: p.
  Includes index.
  1. Power resources.   I.   Title.
TJ163.2.H34   1977   333.7   76–5128
ISBN 0–06–011777–X

77 78 79 80 81 10 9 8 7 6 5 4 3 2 1

Contents

PART ONE

Energy from the Earth

1. The Energy Crisis Is Real

National security, the Nation's economy and the ability
to determine life style are all in peril today. Substantial assistance from new technology is critically needed,
but significant results are not expected before 1985.
Major efforts must be pursued now because of the time
required to research, develop and implement new
energy technologies.

To ensure maximum flexibility for future energy
systems, and to allow for some failures in the development process, the Nation's Energy Plan must provide
multiple options which, taken all together, could exceed
perceived needs.

Accordingly, today's national energy research,
development and demonstration programs must:

Open up new choices for the future

Avoid overemphasis on single approaches which tend
to foreclose future options

Shorten the time for transition to new fuel forms
based on abundant domestic fuel resources.

The task of creating choices for the future must be
urgently addressed now—and with full public
participation.

—ERDA National Energy Plan, June 30, 1975

There have surely been enough books documenting the energy crisis.
It might be possible to let them be prologue to this book, which sets
out an alternative—or rather a number of alternatives—to our traditional energy style.

In this chapter we will examine the validity of the energy crisis
itself, the extent of conventional energy sources, and the prospects for
conservation, or the reduction of population increase or per capita
consumption of energy.

The book will then recount the good news, namely that there are other ways to go about using natural resources that have been part of the Earth since long before man had any need for energy. These homely alternatives include Earth, sea, wind, and sun.

But first, the bad news.

Our fossil fuels of coal, oil, and gas are in a sense stored-up nuclear energy: Solar energy trapped in trees and other vegetation, and in living things, has been converted to carbon over millions of years. And the sun itself is a nuclear power plant. This fortuitous process continues, but at such a leisurely pace that we cannot wait for enough to be produced to match our voracious demands and escalative consumption.

A recent article about oil and gas futures in *Science* begins with this interesting statement: "If the U.S. Geological Survey is right, the United States is at least a decade away from seriously depleting its domestic oil and gas resources." A ten-year supply of oil and gas hardly comes under the heading of encouraging news, but the remainder of the opening paragraph makes even that sound like sheer optimism: "But if several distinguished disbelievers of the Geological Survey are right, the United States is running out of oil and gas right now."

The gist of the article is that while the United States Geological Survey has for years used a method (introduced by one of its researchers, A. D. Zapp) indicating that the continental U.S. has undiscovered petroleum reserves of 200 to 400 billion barrels, and 1,000 to 2,000 trillion cubic feet of natural gas, Mobil Oil Corporation researchers have recently scaled down that optimistic outlook to only 88 billion barrels and 443 trillion cubic feet, less than half of even the most pessimistic USGS predictions. In agreement with Mobil's frightening estimates is a study by researchers at the National Academy of Science. One of them, the noted authority M. King Hubbert, stunned the energy community with his opinion that less than 50 billion barrels remain to be discovered! Furthermore, he pointed out that U.S. oil production peaked five or more years ago and that natural gas is about at that point.

With price controls off for natural gas, it is hoped that prospectors will search out new sources and also that previously unprofitable reserves will be put into production. Similar incentives might also encourage petroleum production. Yet in the last several years oil companies have drilled a total of sixty-five holes off the coast of Newfoundland and only three yielded any oil at all, in each case too little to handle profitably. The cost of this disappointment ran to about $200 million.

Oil shale is often suggested as our salvation, but this bonanza shrinks drastically under close scrutiny and all but disappears when followed through the total process required to squeeze out oil. Federal Energy Administration authorities pointed out in the Project Independence Blueprint that it may cost more in energy to produce oil from shale than the oil is worth. And hell hath few furies greater than those of environmentalists dedicated to protecting shale deposits from such desecration.

Barring large new discoveries, natural gas will be burned up in a matter of years, a decade or two at best. Oil will not last much longer, despite the seeming hoard recently discovered and belatedly being sucked out of the Alaskan tundra. So we are stuck with something as old-fashioned as coal.

After all the bad news, it is good to know that we have an estimated several centuries of coal remaining, at present rates of consumption. This would seem to make it possible to forget our responsibilities to posterity, as it is difficult to work up much sense of duty to descendants removed some twenty-five generations. But there is a catch to the coal bonanza as well, even beyond the problems of technology and environmental impact that stand between all those thick black seams and the consumer's coal bin.

The catch is the qualifier, "at present rates of consumption," of course. For as gas and oil peter out, we must greatly increase our use of coal. Coal has other drawbacks, too. Most of our reserves are soft coal, high in sulfur content. There are many techniques for remedying this shortcoming, including coal gasification and liquefaction. Both techniques have been around a long time, but we have yet to build large-scale plants to produce appreciable quantities of clean coal-derivative fuels. It must also be remembered that it may cost us upward of 40 percent of the energy in coal to convert it to tidier gas or liquid fuels.

President Eisenhower, AEC officials, and legislators at signing of the Atomic Energy Bill in 1954.

There was a time when the nuclear power proponent would interrupt such a recitation of doom and point to his solution. That panacea was promised a quarter of a century ago when the Japanese mushroom clouds at last faded into guilty memories and engineers began to beat that apocalyptic sword into the plowshare of nuclear power. A thimbleful of uranium would drive great ships around the world, we were told. Nuclear fuel would make energy worries as old hat as buggy whips and wars.

Alas, it was a false promise. Having come a quarter-century since that bright promise, and having spent tens of billions of tax dollars along the way in direct subsidies, we have by no means achieved energy independence. For nuclear power plants, pledged to quickly provide 10 percent, then 25 percent, then 50 percent, and finally *all* of our electric power needs, today provide only about 7 percent. Worse yet, this is only about 1 percent of our *total* energy needs.

It must greatly embarrass nuclear people who have committed

their careers and their capital to fission reactors, to realize that solar energy still provides several times as much electricity as do nuclear plants. For hydropower, working at dams such as Grand Coulee, Hoover, and Bonneville, is solar energy but once removed. So is wind power, which electrified much of our nation before TVA and cheap central utilities.

One brief reference to the hazards of nuclear power plants. After several decades of operation, no "civilian" deaths have occurred, despite the hundreds of accidents that have taken place. Although there are many problems with radioactive wastes, there have been no deaths directly attributable to them. And the warnings of Linus Pauling and other concerned scientists may be overly cautious caveats.

There are hazards, and the reader is free to make of them what he will. I will simply note that there may be reason for concern when the federal government itself has to provide the insurance for nuclear plants, and in amounts only a fractional part of the estimated worst

One of the geothermal power plants operated at The Geysers in California. That is water vapor coming from the facility. *Pacific Gas & Electric Co.*

cases arising from a possible nuclear catastrophe. Also, the government has admitted the dangers of sabotage and blackmail.

Let us here consider only the economics of nuclear energy. Although some would have us believe that nuclear power is cheaper than conventional power, this is not so. Even if we discount all the billions of dollars underwritten by the government, the insurance paid by taxpayers, and the additional costs of attempting to store wastes that obviously cannot be disposed of once and for all unless we fire them into the sun, as some nuclear planners have urged.

Those nuclear proponents who once loudly defended the nuclear promise know better than most of us the shortcomings dooming the nuclear approach before it has more than begun to make a dent in our energy woes. There are major problems; we will look at fuel first.

For all its sophistication, the nuclear power plant is simply a heat engine. Surprisingly, it is less efficient than old-fashioned coal, oil, or gas-fired machines (and it cannot hold a candle to the waterwheel). The best conventional power plant is about 38 percent efficient; it wastes 62 percent of the heat content of its fuel. The nuclear plant wastes 70 percent, and thus it also thermally pollutes the environment that much more.

That was the easy part. Here is the hard one. Surprisingly, except to the nuclear expert who has known it from the start, presently operating nuclear plants—called low-pressure water reactors—waste 99 percent of their uranium fuel before even getting to the heat-to-energy conversion that wastes 70 percent of the heat produced! A pound of uranium costs as much as a ton of coal. The beauty of nuclear fuel is that it has far more energy per pound than fossil fuels. But if we waste all but 1 percent of it, the benefits dwindle.

There was a time when nuclear stockpiles got so plentiful that prospectors could no longer make a quick million by climbing into a Jeep and roaming uranium country with a Geiger counter. That time seems past. We still have only a few dozen nuclear plants operating. Some have failed and have been dismantled—at a cost of millions of dollars. A large variety of problems have slowed the promised growth of nuclear plants. Few foreign lands have nuclear power plants. Yet there are warning signs that the seemingly endless supply is definitely limited and may be exhausted for all practical purposes in as little as thirty years.

Adding to the problem of nuclear power is the skyrocketing cost of uranium. Recently priced at about eight dollars per pound, the figure has tripled. While nuclear proponents claim that nuclear power is so cheap that fuel costs are not even a factor, a major manufacturer of reactors is now trying to avoid its contractual commitments to furnish fuel for its commercial power plants. If the host of promised new plants here and abroad are built, what then? At least the environmentalists will not have as much to worry about except freezing in the dark when fuel runs out.

For a long time the "breeder" reactor was hailed as the new nuclear hope. Where the conventional reactor wastes 99 percent of its expensive fuel, the breeder wastes only 30 percent. Here would be a cheering seventy-fold increase in the effectiveness of nuclear fuel, as the breeder "bred" more fuel than it consumed. Alas again for nuclear dreams, the breeder has been plagued for years with a number of problems that now seem insurmountable.

While it is true that breeder reactors are operating in some foreign countries and that some nuclear people claim that we could do the same in the United States but for "politics," it is becoming clear that barring unexpected sudden breakthroughs, the breeder as a safe follow-on to the earlier light-water reactor is still a long way off.

In the summer of 1975, a decade after the catastrophic runaway of the experimental Enrico Fermi breeder reactor, the U.S. Energy Research and Development Administration (ERDA) noted in summing up its program for the future that the breeder, which it had been hoped would be "on the line" by 1985, would not become a factor before the year 2000.

President Richard M. Nixon once answered a complaint that the breeder program was far behind schedule by blithely suggesting that if need be our nation would "leapfrog the breeder"—to fusion power. Fusion is the stupendous nuclear process that goes on within the sun itself. Rather than fissioning, or breaking apart, atoms fuse together, with a byproduct in the form of energy. Proposed and attempted even

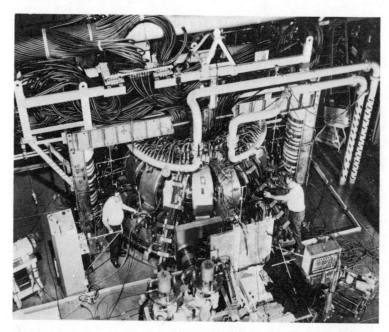

Fusion experts at Princeton's Plasma Physics Laboratory at work on complex equipment of a Tokamak device, inspired by Russian designs. *AEC*

before fission power, the fusion process at this writing has not yet been demonstrated as even scientifically possible, much less technologically or economically feasible.

In the 1950s Dr. Edward Teller, father of the H-bomb or fusion bomb, told me he did not expect to see fusion power commercially successful in his time. In recent years, with work on laser-triggered fusion showing some promise, he has become more optimistic. I hope that he is right, but we cannot hold our breath until success comes. For fiscal 1976, ERDA has budgeted $300 million for fusion research. Surely we must pursue all possibilities, especially one with such bright promise as the fusion power plant. Operating on fuel produced cheaply from sea water, it will produce power cheaply, practically forever, and with no danger. Or so the story goes, sounding remarkably like the great atom-power promise of the mid-1940s. But even if it should prove possible, we will still be fifty years or more from commercial power plants. What do we do for power in the meantime?

Many solutions have been proposed to our energy problems, and there is the ring of hope in many. By conserving, we could doubtless save an appreciable amount of energy. By holding the population at present levels, and by continuing such approaches as economy cars and better building insulation, we can stretch our limited resources a bit longer. There are more ambitious suggestions as well.

In recent years the vision of a "hydrogen economy" has been mentioned with increasing frequency. For hydrogen, the engineers tell us, is the perfect fuel, if there is such a paragon among resources. Hydrogen has a high-energy content for its weight and volume; it burns with great efficiency and minimal pollution of the environment. It does indeed seem to live up to its glamorous billing. The catch is that hydrogen does not grow on trees. To be sure, there is plenty of it in water and in other places as well. But getting it out of such chemical combination generally takes as much energy as the hydrogen will then give us back. Until someone finds a pure hydrogen farm, or mine, or lake, the hydrogen dream is little more than that.

There are other energy visions, including the old standby of perpetual motion. Inventors still delude themselves (but not the Patent Office) with falling weights, magnets and shields, and frantically spinning counterweights said to produce power from nothingness. None of these wonderful inventions has yet gotten off the ground, and it is impossible to believe that the oil companies have succeeded in suppressing all of them.

Electrical wizard Nicola Tesla experimented with esoteric energy sources, and partisans of his claim that we can somehow tap great reservoirs of electromagnetic energy in the atmosphere. Others say we must go to the great radiation belts surrounding Earth, or far beyond to the realm of "astral power," "antimatter," or "Dyson stars." Most of these schemes make nuclear fusion seem like rubbing two sticks together to make fire, and there seems little hope of any help from them in time to save us—or our tenth-generation descendants.

Humanity has reached many junctions and crossroads in its long journey from the cave to the electrified megalopolis. Each step and each jump have been toward increasing technology and more complex schemes of producing and using power. Now, ironically, at the seeming peak of our powers we may have to accomplish what amounts to a dramatic and humbling U-turn if we are to continue to progress— or even to run in place.

We first made use of fossil fuels quite recently, and until human beings began to burn coal and petroleum, they relied on a completely different kind of energy: natural energy *in real time*. Burning coal and burning wood seem much alike, but there is a tremendous difference between the two processes: Coal is not replaceable; wood is. Grinding grain with a waterwheel seems far less elegant than doing the same thing with a steam engine or electric power. But water is not consumed, as are coal, gas, and uranium.

Up to the time we dug into Earth for the stored-up treasure of eons, mankind depended on nature not just for food, light, and heat but also for power. Sunshine provided us with food and fiber and warmth. Wind drove our ships and our mills; water ran the engines. The single fuel we burned produced only a nuisance type of smoke, and with proper management would last as long as the sun shone on forests to make more trees. Under this natural economy, civilization could have thrived indefinitely, except as increasing population overloaded the system.

As luck would have it, those quaint old energy sources are still here and all we have to do is learn once again how to use them. A bit humbly we are turning back to what are now generally called alternative energy sources. Most people think of them as supplements rather than substitutes; a temporary stockpile to give us time to get conventional sources back on the track. But there are those of us who think that it is possible to drive the world on income energy; safe, abundant, and everlasting supplies from the earth, the water, the wind, and the sun themselves.

2. Geothermal Energy

In Hell there is a place called Malebolge,
Made, like the cliff that circles it around,
Entirely of stone, and iron-colored.
Right in the center of this evil field
There gapes a very deep and spacious well,
Whose form I shall describe in its due place.

—Dante, Hell

It has often been suggested that Dante's vivid vision of hell was conjured up by visits to the strange Lardarello area, where sulfurous fumaroles, mudpots, craters, and boiling wells splotched the land. This outpouring of geothermal energy was indeed triggered by molten rocks in the bowels of Earth.

The hellish connotation of geothermal wells has been softened since Dante's time, and most of us are more likely to think of the geyser Old Faithful at Yellowstone Park. Less known is The Geysers, a geothermal site in northern California, subterranean steam producing some 400 megawatts of electricity for the power lines of the Pacific Gas & Electric Company, more than half the needs of San Francisco. Here is the reason for the keen interest suddenly focused on this long-known but little-used alternative energy source.

All of our fuels spring from the Earth. Wood and other combustibles have their roots in the soil. When demand exceeded the available wood supply, miners dug up coal and later on drillers probed for petroleum and natural gas. It is not remarkable, then, that there is another treasure in energy far beneath the surface. Not fuel to burn to produce heat, but heat itself. Yet most of us are amazed to learn that geothermal energy has been tapped for seventy years as a source of electricity. Worldwide, there are now an estimated 1,800 megawatts of power produced by natural steam wells in California, New Zealand, Mexico, El Salvador, Iceland, Russia, Italy, and Japan. Optimistic proponents of this alternative energy source project a total output of up to 20,000 megawatts in 1985, and perhaps as much as 400,000 megawatts by the year 2000; 40 percent of total electric needs at that time.

Geothermal's Past

The most obvious indicators of geothermal activity are the hot springs that flow abundantly on Earth's surface. These steamy hints of what lies deeper have been known for ages and have been put to good use for the last two thousand or more years. Hot water is basically useful for bathing and washing clothes. In time, people came to attribute medicinal value to hot springs or "spas." The ancient Greeks, Romans, and Babylonians, and the Japanese are among those who knew and used hot springs for curative as well as recreational purposes.

The Romans developed hot springs throughout their empire, as far north as Bath, England. Later the Hapsburgs established the famous Marienbad and Karlsbad spas in Czechoslovakia. More than a century ago Hungarians went nature one better and *drilled* a well to reach geothermal water. "Taking the waters" has long been a way of life, and President Franklin D. Roosevelt helped popularize Warm Springs, Georgia, to whose mineral pools he went often to bathe.

Dante may have equated Lardarello's vaporous springs with hell, but the Icelandic concept of hell is a frozen horror of ice and snow. To the Icelanders the phenomenon of boiling hot water from the cold ground must have seemed instead a miracle from heaven. All around the world human beings have used geothermal waters for a variety of purposes, including bathing, washing, agriculture, cooking, and for the recovery of byproducts like boric acid, sulfur, and other chemicals. At first, geothermal energy was roughly the reverse of lake ice cut into blocks and packed in sawdust for its cooling magic in summer. But in 1904 a new age of geothermal energy began when engineers tapped the steam from Lardarello to run a $\frac{3}{4}$-horsepower generator and light five light bulbs. It was a tiny effort, but it proved a principle: heat from deep within the Earth could run an electric power plant. Today the output from Lardarello has grown from five light bulbs to 400 megawatts of power, enough to fill the electric needs of about half a million people.

The Waters under the Earth

It has been pointed out that the Earth was born hot and still has not cooled down. While the crust we live on is quite comfortable most of the time, the temperature increases as one bores down through that crust. As miners know, the increase is something like 100 degrees F. per mile. At the base of the "continental crust," the temperature may be as high as 1,800 degrees F., and at Earth's center something like 8,000 degrees F. Should we ever succeed in running a pipe (of material able to withstand such temperature) into that hot furnace, we could draw on a fantastic reservoir of heat energy. The fact that we now dig no deeper than six miles or so lessens the geothermal potential, but even so it is remarkable.

Experts believe that most geothermal heat is caused by the decay of nuclear materials deep in the Earth, with some heat coming from the friction of rock movement, tidal forces, and perhaps other sources. Whatever the causes, it is a fact that beneath the crust lie great quantities of molten rock, or "magma." Occasionally this magma breaks through to push up as lava in volcanoes. Not as spectacularly, but more regularly, geothermal heat from magma reaches the surface as hot water. In a few places it does this in a spectacular manner as highly pressurized hot water is vented to the air and turns to steam.

Two ingredients are necessary for the production of geothermal wells: magma, or hot rocks, and a source of water. There is discussion whether this is surface water, or new water produced from elemental materials. Whatever the source, water heated by magma in certain areas is available for a variety of tasks, ranging from fish farming to generating electricity.

Geothermal energy is available in four forms: dry steam, wet steam, hot rocks, and geopressured deposits. Dry steam is ideal for

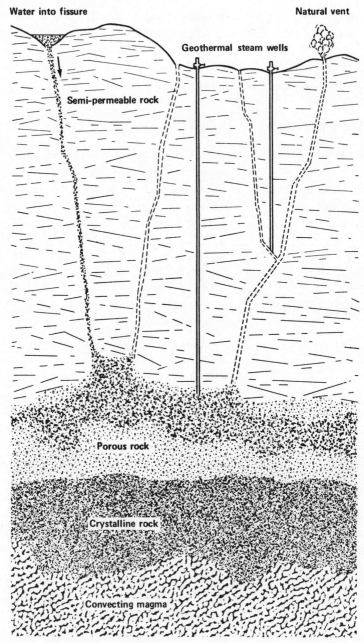

Water into fissure

Natural vent

Geothermal steam wells

Semi-permeable rock

Porous rock

Crystalline rock

Convecting magma

Heat from the magma at the bottom of this drawing converts ground water to steam or hot water. This is the source of geothermal energy. *ERDA*

producing power, since it causes no corrosion of turbines or associated equipment and is the most potent heat source because of its very
high temperature. Wet steam is also a useful source but must be
cleaned up so that it won't harm generating equipment. Hot rocks are
tantalizing geothermal sources, but water must be introduced artifi-
cially to the dry heat source to produce steam or hot water.

Very high-temperature geothermal steam is found only in regions
of youthful volcanism, earthquake faulting, or recent mountain building. In the United States, this is the western part of the country. The
remainder, from the Rocky Mountains east, seems to have only low-
temperature geothermal energy, and this is particularly true of the
Gulf Coast geopressured zones. Other low-temperature geothermal
energy is found in Western Siberia, in portions of central Europe
north of the Alps, and in the Carpathian Mountains.

Geopressured deposits have been trapped for ages by clay beds and
are much hotter than normal underground water. Such deposits occur
even in areas where there are no volcanoes, earthquake faulting, or
mountain building. In the United States the largest area of geopressured water is along the Gulf Coast. There are two unusual things
about such geothermal deposits. First, the water is very low in salinity, since it has been effectively filtered by clay and perhaps by shale
as well. Second, the water may also contain appreciable amounts of
natural gas—up to a cubic meter of gas per barrel of water. Thus
there is a double reason for seeking such resources.

The Gulf Coast zone is part of a geopressured area perhaps several
thousand miles long. In width it extends about 100 miles inland and
150 miles out into the continental shelf. Smaller geopressured zones
have been found in California, Colorado, Oklahoma, Utah, and
Wyoming. Foreign zones include those in Mexico, South America, the
Middle East and Far East, Europe, Africa, and Russia.

Putting Geothermal Energy to Work

Primitive use of geothermal springs is many centuries old, as we have
noted. A century and a half ago some people were beginning to
exploit this source in other ways. Wells were drilled for hot water,
and chemicals were reclaimed from hot springs, notably at Lardarello

where a thriving boric acid industry sprang up. By 1900 hot-water drilling was common in Italy, Hungary, Germany, and Iceland. Budapest accomplished geothermal home heating in the 1930s, and by the early 1940s Reykjavik, Iceland, was using geothermal hot water for home heating and industry. New Zealand, notably at Rotorua, was doing the same thing. Japanese farmers heated greenhouses with hot springs shortly after the turn of the century.

Russia too has heated homes for a long time with geothermal springs. More recently Russia's engineers have gotten into power generation. So have engineers in Mexico and El Salvador. Other

Iceland, surprisingly, is a leader in geothermal exploitation. Steam wells like this one provide heat for Reykjavik, the capital city. *ERDA*

interested nations include Ethiopia, Kenya, the Philippines, Indonesia, Chile, Turkey, Katanga, Zaire, Nicaragua, and France.

The American Experience

With its more than one thousand geothermal springs, the United States has known for a long time about geothermal energy. Yellowstone, protected by law against exploitation along with geothermal fields at Katmai and Lassen National Parks, has long demonstrated the tremendous power pushing its way through Earth's crust. Test drilling was done at Yellowstone in the 1920s and 1930s; in 1967 and 1969 more sophisticated prospecting turned up 240-degree C. water at a depth of only 300 meters.*

California, site of the great gold rush, was also a leader in geothermal exploitation. In the 1920s pioneers bored wells at Niland, near the Salton Sea. There was plenty of low-temperature steam, but the project was abandoned because of a lack of a market for this commodity. At the northeast end of the Niland field shallow wells produced carbon dioxide, however, and this gas did have commercial value. From the early 1930s through the mid-1950s some sixty-five wells were drilled, with about 100 million cubic meters of carbon dioxide produced to make dry ice for refrigerator cars.

While the Niland pioneers were working the southern part of the state, prospectors northeast of San Francisco hit steam in the area logically named The Geysers. The place was discovered by explorer William Bell Elliott in 1847 when he blundered into a canyon where steam poured from fissures along a quarter-mile length. Like Dante in an earlier time and halfway around the world, Elliott suspected he had trespassed the gates of hell.

In 1922 drillers attempted to harness The Geysers for electric power but succeeded only in lighting the hot springs resort which had sprung up. Steam corroded the engines and plumbing, and there was plenty of hydroelectric power in the neighborhood. In short, the United States wasn't ready for geothermal power, and would not be for several decades. For years the extent of geothermal applications was for agriculture, fish farming, bathing, and house heating as at Boise, Idaho, and Klamath Falls, Oregon, and use of hot water in

* A table of metric equivalents appears on page 181.

explosives manufacture at Steamboat Springs, Nevada. But the potential was too great, and in 1956 a second wave of engineers tackled The Geysers.

Magma Power Company and Thermal Power Company drilled wells in The Geysers area and began producing dry steam. In 1960 Pacific Gas & Electric contracted with them to buy the steam, and in short order caught up with and surpassed the output of Italy's impressive Lardarello geothermal plants.

A new day was also dawning in Imperial Valley for geothermal energy. Although an experimental 3-megawatt power plant at Niland in 1960 was unsuccessful, the Imperial Valley area soon became one of the most exciting geothermal prospects. Standard Oil of California drilled a well west of the town of Brawley in 1963 and found brine at 260 degrees C. Deeper down the temperature went as high as 370 degrees C., and according to scientist Robert Rex the 90-mile-long geothermal bed contains a billion acre-feet of water or more, a tremendous source of both power and fresh water. Magma, which helped produce the successful well at The Geysers, also began drilling in the Imperial Valley area. The U.S. Office of Saline Water and the U.S. Bureau of Reclamation pushed programs for geothermal energy, and recently the San Diego Gas and Electric Company signed a contract for a geothermal plant at Niland, site of the unsuccessful prospecting more than half a century earlier.

How Much Energy?

To give an idea of geothermal potential, 40 cubic miles of geothermally heated rock is estimated to match the energy in all of Alaska's North Slope oil find. And 40 cubic miles is a chunk little more than $2\frac{1}{2}$ miles square and 6 miles deep, about as deep as we can successfully drill for wells of one kind or another.

Estimates of geothermal potential depend on which expert one listens to. A few years ago the National Petroleum Council reported that by 1985 the United States could develop between 1,900 and 3,500 megawatts of geothermal electricity. For comparison, the higher estimate is about half enough electricity to supply New York City. This estimate was very low compared with that made by the

Hickel Conference, chaired by Walter Hickel, former Secretary of the Interior. The Hickel group estimated geothermal potential by 1985 at 132,000 megawatts, about forty times as much as the National Petroleum Council estimate. Many other estimates have been made and these range from an encouraging 20 percent of our electrical needs by 1985 to only one-half of one percent.

The Hickel estimate seems too optimistic as a matter of fact. Since geothermal electric plants tend to be rather small (about 5 megawatts per well on average), the amount of power predicted would require the establishment of 26,000 producing geothermal steam wells by 1985. Since not all wells produce, perhaps 50,000 would have to be drilled, at a cost of about $10 billion between now and 1985. It is doubtful that this will, or even can, be done. In the last five years only about 150 geothermal wells are estimated to have been drilled. For comparison, approximately 28,000 oil wells were drilled in each of those years. The number of wells needed to provide 20 percent of our electric power seems possible but somewhat difficult to achieve.

The U.S. Geological Survey has designated about 1.8 million acres in the western United States as "Known Geothermal Resource Areas," or KGRAs. An additional 96 million acres are designated as "Prospective Geothermal Resource Areas." Even for a KGRA, this does not mean that a producing well has been drilled but that the geology and other indications suggest the area is encouraging enough to be developed.

The western United States includes 948 known hot spring areas. The rest of the continental U.S. plus Alaska and Hawaii have an additional 165, making a total of 1,113. There are a reported 5,231 geothermal springs in the entire world, giving our country about 20 percent of the total. Surprisingly, Iceland is second with 516. Japan has 298, India 202, the USSR 140, Java 155, Italy 149, France 124, and Turkey 113. The least fortunate country is Sweden, not listing a single geothermal well.

To put the geothermal potential into better perspective, it has been estimated that if all the heat that is stored in the Earth's crust down to a depth of six miles could be put to use in heating our homes, every American could live in a geothermal heated residence. Unfortunately, the realistic prospects are not all that rosy. Although it is well agreed that the heat energy in the outer six miles of Earth's crust may be 2,000 times greater than that of all the world's coal, making use of it is something else again. Best guesses are that we may never recover more than a fraction of one percent of that tantalizing total of geothermal heat. But even $\frac{1}{1000}$ would be double the coal stockpile!

Electric Power

We live in an increasingly electrified world, and it is the power production capabilities of geothermal energy that command the attention of governments and scientific communities alike. With a track record of sixty-five years of commercial electric power production at Lardarello, geothermal technology is well past the experimental stages. Four sizable geothermal electric complexes are operating in the world today: Lardarello, Italy; The Geysers, California; Cerro Prieto, Mexico; and a large area in New Zealand.

Primitive peoples used fumaroles for cooking and the condensed steam for drinking water. Later, sulfur, kaolinitic clays, and mercury and alum were reclaimed from the water. Boric acid reclamation began at Lardarello in 1812. Wood fires were first used to boil out the boric acid, then fumarolic steam was substituted in 1827. Wells were also drilled for steam. By 1904 electricity was produced experimentally and in 1913 a 250-kilowatt plant came on line. This was the beginning of continuous generation of geothermal electricity at Lardarello.

In Italy, the boric acid recovery works shut down in 1969 because it could not compete with other sources of borax. But by then the power-generating capacity of the wells had grown to nearly 400 megawatts. Lardarello supplies 365. Some 25 come from Monte Amiata, seventy-five kilometers southeast. Generator sizes range from 99 kilowatts to 26 megawatts. These are noncondensing types, but there are plans to switch to the much more efficient condensing turbines.

The total geothermal field area at Lardarello is in excess of 250 square kilometers. There are also other fields east and southeast. About 500 wells have been drilled, of which about 200 are still producing. The average well depth is slightly more than 1,000 meters.

Geothermal power plants at Lardarello, Italy, are the oldest in the world, and make up one of the two dry-steam operations in the world. *Scientific American*

Infrared imagery is now being used to find anything overlooked. No adverse effects on agriculture from borated effluents have been reported. At Lardarello, most of the land is given over to farms, orchards, and vineyards.

Since the first 12.5-megawatt plant went on line at The Geysers in 1960, others have joined it at regular intervals until today there are ten plants producing a total of about 400 megawatts. The plants will soon provide 700 megawatts, sufficient electric power for the city of San Francisco, and projections for 1980 are 1,200 megawatts, the size of a very large nuclear power plant.

The Geysers is dry steam, like Lardarello. Wells range from 4,000 to 8,000 feet deep and produce steam at a temperature of 300 degrees C. The steam is filtered before passing through the turbines to remove any particles that might damage equipment. Condensed steam is evaporated in cooling towers and allowed to escape to the atmosphere, with only a small fraction of it reinjected into the ground. Carbon dioxide is the principal gas byproduct. At one time it was thought that all the dry steam near The Geysers site had been located, but a few years ago Geothermal Kinetics acquired adjacent land and has since produced dry steam from several wells.

Unfortunately, there are not many such dry-steam sites and power production using other sources has not been crowned with the brilliant success of the Lardarello and The Geysers installations. In Imperial Valley, for instance, although a huge hot-brine deposit was discovered years ago, there is not yet a successful, large-scale power plant in operation. The brine itself is the biggest problem. It cannot be run through the turbine equipment since it would soon corrode and destroy the equipment. Nor can it merely be dumped on the surface after use—pollution of this kind cannot be tolerated for a variety of reasons. Imperial Valley is one of the lushest and most productive farming areas in the world and the farmers want to keep it that way.

Despite early work in California around the Imperial Valley geothermal resource, it was Mexico that first exploited the region's hot brine for electric power production. That country's pioneer power plant was built in the Mexicali Valley of Baja California, near the Cerro Prieto volcano. Here is an area of obvious geothermal activity: active small volcanic craters, steam spouts, geysers, and mudpots.

The Mexican government is putting the energy that lies beneath to good use. The first of two turbogenerators, each with a 37.5-megawatt capacity, began producing electricity in April 1973. In September the second unit went on line, and both have been active ever since. The plants use wet steam from wells of a maximum depth of 1,600 meters. The water is far less contaminated than that to the north in California and has a salt content of only about 2 to 3 percent. Special alloys are used in the turbines, and the equipment has not been damaged by the salts.

The Cerro Prieto geothermal plant is so successful that plans call for adding 55-megawatt units more each year beginning in 1977. This will give a total output of 350 megawatts by 1981, making the facility a sizable electric power plant. Meantime, the Mexican Federal Power Commission is surveying many other sites, and at least four are in various stages of development.

New Zealand has a very large geothermal region, some 150 miles long and 30 miles wide, stretching between active volcanoes in the

Mèxico's pioneer geothermal power plant at Cerro Prieto, Baja California. This is one of seventeen producing wells. *U.S. Dept. of Interior*

New Zealand's Wairakei-Broadlands geothermal facility. Both steam and hot water are produced by these wells. *ERDA*

center of North Island to another active volcano on White Island. The Maoris have historically used hot springs for cooking, washing clothes, and bathing. European settlers later installed primitive equipment to provide hot water for a variety of uses. But it was not until after World War II that the New Zealanders had to look for new sources of power, and tapped geothermal heat as a likely prospect.

Drilling started at Wairakei in 1950 and by 1960 commercial production of electricity began. The New Zealand wells produce hot water and steam. The steam is separated from the hot water and fed to the power station. The hot water is also collected and pumped at high pressure into "flash tanks" which produce lower temperature steam still good for power generation. Total capacity of the Wairakei-Broadlands field is about 170 megawatts. Ironically, further development of New Zealand's geothermal resource has been slowed by the recent discovery of large natural-gas deposits.

Japan was among the pioneers in geothermal electricity, having begun experiments in 1919 on Kyushu Island. By 1924 a pilot one-kilowatt plant was in operation. At Matsuwaka drillers found dry steam in recent years and this has been used to power a commercial plant producing 20 megawatts. There is also a 13-megawatt plant in the Otake field on Kyushu Island. Mitsubishi Metal Corporation operates an electrolytic zinc plant at Akita and geothermal steam wells with a potential of slightly less than 10 megawatts will provide about 20 percent of the electricity needed to operate the plant.

Following the discovery of geysers in the southeastern part of the Kamchatka Peninsula in the 1940s, the Russians began development of geothermal electric plants. Commercial production got underway in 1967, and presently is in the range of 30 megawatts, with costs said to be 70 percent of conventional costs. At Paratunka the Russians operate the first geothermal electric plant using Freon rather than steam as its working medium. The Freon is heated by water at about 80 degrees C. and drives the turbogenerator equipment in what is called a binary system.

Iceland is at the lower end of the geothermal electric producers. Most of that country's applications make use of heat, but there is a 3-megawatt electric power plant operating in the Mamafjall geothermal field.

Dry steam is the power engineers' dream. It is easy to handle, does not corrode equipment, and produces more power than lower temperature wet steam or hot water. However, this does not mean that there is little potential in wet-steam geothermal fields. New Zealand's Wairakei plant demonstrates that mixtures of steam and hot water can be handled and that hot water can be changed to steam.

There are fascinating possibilities of using part of the hot water, at lower temperatures, for desalination. Picture a large, multipurpose geothermal facility, with as much steam flashed off as possible to run the electric plant, waste hot water then desalted, and the remaining brine picked clean of its contaminants, many of which are valuable minerals or chemicals! Thus a geothermal resource that might not be economical for electric power alone, or for desalination or recovery of minerals and other constituents, might succeed if all three things were done at the same site.

This approach may be taken in the Imperial Valley resource. Underlying the desert is an estimated billion acre-feet of water, and water is vital to that dry land. It has been shown that a desalter in conjunction with a geothermal power plant might produce half as much fresh water as now comes from the Colorado River.

Putting the Heat to Work

While electric power is the glamor application of geothermal energy, and the all-electric home is the goal of most planners, this is not necessarily the best way to go. Electricity does not grow on trees, but must be generated by running a heat engine. In the process of changing heat to kilowatts, most of the heat is wasted, unfortunately. We have noted that the best fossil fuel power plants waste more than 60 percent; nuclear plants do even worse. Geothermal electricity, because it is produced by relatively low-temperature steam, is the least efficient kind. In fact, it is generally less than half as efficient as a large coal-fired plant.

All of which suggests that for heating applications it would be wise to use the steam or hot water as it comes out of the ground, rather than going through the wasteful conversion process twice: once from heat to electricity, and again from electricity back to heat.

Geothermal wells did heating chores long before electricity was available, and they are still doing them and other chores as well, including refrigeration, for this can be done with hot water.

Just as electric power plants must be sited close to the geothermal wells they are to tap, a community or industrial heating system must be near its geothermal source of supply. So we probably will not pipe steam or hot water great distances from a geothermal well. It will be only those close to a hot spot who will benefit—unless we learn someday how to transport steam without heat losses.

In what has to be a fluke of nature, Icelanders have long made great use of geothermal wells. Nearly half the population of the country lives in houses centrally heated by natural hot water. Reykjavik is almost totally heated in this manner, with a municipal system supplying about 85,000 people. Plans call for serving about 65 percent of the total population of Iceland before 1980. Geothermal heat costs the Icelanders about 60 percent of what oil heat would cost, and pollution is greatly reduced. Greenhouses are also heated by geothermal wells, and an estimated 20,000 tons of fuel are saved this way. Natural hot water is also used in fish-breeding stations. An estimate of fuel oil saved in the total geothermal system runs to 360,000 tons per year.

Russia gets a large percentage of its heat from geothermal wells, and since it is such a huge country the total consumption is far greater than in tiny Iceland. Several towns use domestic and industrial hot water from geothermal wells. Refrigeration is provided in some cases, greenhouses are heated, soil is thawed, and swimming pools and baths are warmed. Russian scientists have estimated a geothermal potential for the Soviet Union of 22 million cubic meters of hot water and 430,000 tons of steam daily. If all this energy could be harnessed, it would save 40 million tons of fossil fuels a year.

The Japanese heated greenhouses prior to 1919. They love therapeutic hot springs and have set aside many geothermal areas as national parks, spas, and baths. At Rotorua, New Zealand, 1,000 wells heat a city of 30,000, and air conditioning is supplied to a 100-room hotel. Construction costs match conventional costs, but operating costs are only 5 percent as high.

Boise, Idaho, has for a long time heated some homes with geothermal energy, and recent discoveries of new sites, including some with dry steam, suggest a bright future for the state. Klamath Falls, Oregon, has some 350 wells heating buildings. There are others elsewhere in the state.

The Hot-Water Prospectors

The first geothermal resources gave themselves away by their hot springs. But there are other prospecting methods being put to use since not all areas have such handy indicators. On-site inspection of a region's geology, seismic or explosive techniques like those used in oil prospecting, and aerial photography, particularly infrared techniques, are being used. "Remote sensing" of regions from high-flying aircraft, and even satellites, offers valuable information as well. Infrared photography or scanning works because it shows up hot spots. To eliminate the normal radiation of sunlight from the Earth, infrared scanning is generally done at night so that only the geothermal areas will show up.

The United Nations in recent years has taken a keen interest in geothermal development. Among its efforts was the use of satellite infrared photos for seeking out geothermal resources in Ethiopia and Kenya. Such an approach is open to nearly everyone since remote sensing satellites like Landsat provide a wealth of photo and scanner coverage of much of the world. Prints cost a few dollars and are available to anyone.

More sophisticated prospecting techniques include electrical resistivity surveys which may differentiate between hot water and cold water underlying an area. Gravitational and magnetic properties of an area may give clues to geothermal "anomalies" or hot spots. "Well logging," the accurate measurement of temperatures and other factors at various depths as a well is drilled, is a valuable technique too. Some geothermal experts claim they can spot areas simply from looking at high-altitude photographs, and point to geothermal steam wells that paid off.

Once a location is selected, drilling is done in much the manner of oil drilling, and holes can be drilled to depths approaching four miles. Geothermal drilling differs mainly in the higher temperatures that may be encountered, necessitating higher temperature materials in equipment. Wells at The Geysers cost about $150,000 per kilometer; those in Imperial Valley somewhat less.

The Hot-Rock Concept

Near Marysville, Montana, a huge area of dry hot rocks has been discovered. Here is a very attractive potential geothermal source—"attractive" because it is believed that there must be many more such sites and that they surpass the steam- and hot-water resources of geothermal energy; "potential" because water must be added to make it useful.

Engineers are working on methods for tapping the energy of hot-rock deposits, again calling on petroleum techniques. In drilling for oil, "hydrofracturing" is sometimes resorted to. In the geothermal application, high-pressure water would be pumped into a dry hole to break up the hot rock. Water would then be pumped down into the rock, heated, and forced back up to the surface for use, as with a natural geothermal well. Scientists at the ERDA laboratories in New Mexico are working on this concept. Others there are working on an even more surprising—and controversial—approach to manmade geothermal energy.

For some time, nuclear experts have been pushing a power-production idea called Pacer. Pacer, in short, is manmade geothermal energy. We have noted that it may be nuclear energy that heats up geothermal steam or hot water; Pacer would use controlled nuclear-fusion explosions to do the same thing. Fusion-power research has intrigued scientists in the United States and in a number of foreign countries, including Russia, for a quarter of a century. While some progress has been made in various laboratories, the successful operation of a fusion reactor seems many decades away. So Pacer is offered as a quick and easy version of fusion that can be developed in a relatively short time.

The beauty of Pacer, in the eyes of its proponents, is that it is not dependent on nature's providing steam or hot water. However, it will be necessary to use underground salt domes where nature provides these handy boilers for containing both steam and the byproducts of

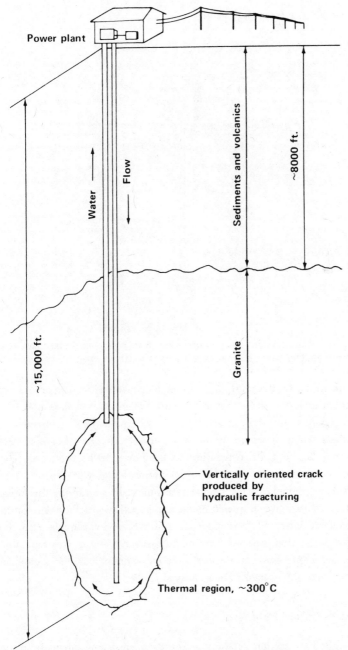

Power plant

Water →

Flow ↓

Sediments and volcanics

~8000 ft.

Granite

~15,000 ft.

Vertically oriented crack
produced by
hydraulic fracturing

Thermal region, ~300°C

A manmade geothermal well. Explosive-created cavity heats water to
steam. Large areas of hot rocks are believed to exist. *ERDA*

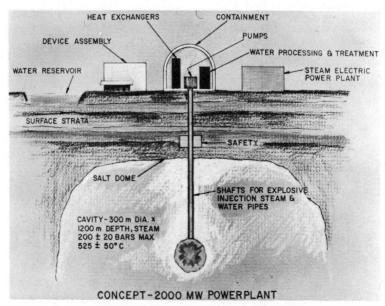

How the Pacer concept would work. Heat from a nuclear blast takes the place of geothermal heat. *Lawrence Livermore Laboratories*

fusion. A carefully controlled fusion blast would be triggered in a suitable underground cavity at a depth sufficient to isolate both the seismic shock and the radioactivity. Water would then be injected to produce steam. It would be necessary to set off fusion blasts at certain regular intervals, depending on the power to be produced. The usual ballpark figure quoted is a 10-kiloton blast every ten hours.

Not surprisingly, Pacer sends environmentalists climbing the walls. Who wants to live atop manmade and repetitive earthquakes or to risk the oozing out of dangerous radioactive byproducts into the steam-production process? Despite all assurances, there are reputable nuclear people who say we don't yet know enough about geological structure to safely set off Pacer bombs.

The Pollution Problem

It is often stated that geothermal energy is clean and harmless to the environment. The question is: compared to what? Certainly such a facility would seem less a polluter than a coal-fired power plant and less hazardous than a nuclear plant. Lardarello has operated safely for some seventy years, and farming operations have successfully coexisted with the power plants. The Geysers' impact on the environment includes clouds of steam, some noise, and a few air and water pollutants. But there are many things to consider when we are making up an environmental impact score sheet for geothermal energy.

The Geysers represents about 15,000 acres of land set aside for power production, compared with about 100 acres necessary for a comparable nuclear plant. Wells must be drilled, roads and plants constructed, transmission lines installed. Waste steam or water must be disposed of, either into the atmosphere or onto the surface, or reinjected into the ground. Wildlife might be affected by any or all of these.

A little thought will indicate that a geothermal plant heats up the surrounding environment, since it releases heat from inside the Earth faster than would otherwise be the case. There are also contaminants in the waste steam or water. Along with pure water or steam are carbon dioxide, carbon monoxide, argon, nitrogen, hydrogen, methane, hydrogen sulfide (which smells terrible), ammonia, radon (a radioactive material), mercury, boric acid, phosphoric acid, boron, arsenic, and fluorides. And fog might be caused by escaping steam, to pose another environmental problem.

Mexico's Cerro Prieto geothermal power plant is in an area where it is possible to drain waste water into surface ponds. But in most areas the salty waste cannot be tolerated and must somehow be disposed of. For another thing, all that hot water coming out of the Earth must be leaving some empty spaces behind, and these may cause subsidence or eventual cave-ins. So in many areas waste fluids must be reinjected. Except that earthquake people point out that this may trigger more frequent earthquakes!

Uncle Sam's Hot Springs

How little interest our country has had in alternative energy sources can be seen in the fact that it was 1963 before the first Congressional hearings were held on geothermal energy. And when legislation was passed by Congress in 1966, President Lyndon Johnson vetoed it because it gave leasing preference to those who had done exploration or development of geothermal resources.

Not until 1970 did the Geothermal Steam Act win passage through the President's signature. In 1974 a follow-up bill, the Geothermal Demonstration Act, called for the leasing of federal lands for geothermal prospecting and development. However, most developers blame the government for being unrealistic in its leasing program. Complaints range from charges that the government itself wants to control and operate geothermal power, to the more easily proved one that it is all but impossible to lease any government land for geothermal investigation. Only on private land has much prospecting been done, and many fear that this situation is likely to continue.

The creation of the Energy Research and Development Administration (ERDA) in January of 1975 may aid geothermal development since a division of the ERDA is charged with that responsibility. Part of the problem has been the involvement of the National Science Foundation, the Bureau of Reclamation, the U.S. Geological Survey, the Atomic Energy Commission, the Bureau of Mines, the Bureau of Land Management, and the Office of Saline Water. Too many cooks, and no central guidance, may have spoiled the geothermal broth for all the many good intentions.

Also complicating the geothermal problem is the law itself. Water law is the most complex ever developed, and geothermal wells represent a special and very confusing kind of water. A key question, as yet unanswered, is who really owns geothermal resources? Do they go with mineral rights reserved by the states or to the federal government? And how should geothermal wells be operated to minimize depletion of the heat resource?

In the western United States, water rights are on an appropriative basis; that is, whoever first developed and used them is entitled to them. In the East, riparian law pertains, meaning that one must own the land through which the water runs. Petroleum law would seem to be a guideline, and the "rule of capture" governs here. This means that one may exploit a well he has drilled even though others are drawing on the same basin or pool. Some common-sense safeguards have lately been added to this concept so that a greedy operator cannot put everyone else out of business. Such adjustments may be needed in geothermal law as development proceeds.

Another vexing problem is how to tax geothermal energy. Operators of The Geysers are allowed by the Internal Revenue Service to classify the steam as a gas, and thus claim a depletion allowance such as petroleum producers are entitled to. Geothermal law is still wide open; surely the states and the federal government will not sit by and watch natural resources being sold at a profit without exacting their due.

How Hot an Energy Source?

Geothermal energy is a fact of life, and present output, while modest compared with large conventional power plants, is a tidy amount of electricity. The several plants operating seem not to have devastated the landscape, nor to be running out of steam or hot water. Indeed, Lardarello is producing as much steam as ever, and some experts say the flow might even increase with time. Geothermal energy, though not inexhaustible, may prove at least a replaceable source. Although continued use of hot water will deplete the supply, given time the reservoir may replenish the supply, much as a farmer's field regains its strength if allowed to lie idle for a certain period of time. Time will tell us if the analogy holds.

We will not produce electricity from geothermal wells except in places where nature provides ample steam, just as we won't operate huge wind turbines in gentle breezes. But for some areas of the United States and many foreign countries, geothermal energy looks like a winner. How fast it develops here depends on the availability and price of fossil and nuclear fuels, environmental attitudes, and the speed with which bureaucracy can move toward opening up geothermal resource areas as pledged in the Geothermal Demonstration Act of 1974.

The challenge is immense. For all the impressiveness of the multi-megawatt geothermal-electric plant at The Geysers, this represents only about $\frac{1}{1000}$ of our electric demand. Filling some appreciable portion of that demand will take much effort, but the reward may be well worth it.

PART TWO

Energy from Water

3. Waterpower

Because of our long use of waterpower, even our language has been affected. Words like millstone and millpond are commonplace, and we sing—or sang—"Down by the Old Millstream." Paris's Moulin Rouge memorialized familiar red mills. *Cockpit (cog pit)*, *fantail*, *governor*, *miller*, and *turbine* are some of the words that stemmed from waterpower technology. And expressions like "water over the dam" and Shakespeare's "More water glideth by the mill than wots the miller of," add sparkle and charm to our language.

Water, even more than food, is vital to life. Like the Earth's surface, our bodies are mostly water. We drink water, cook with it, bathe in it, and cool our drinks with it in frozen form. As steam, it runs our engines, and it is also known as the universal solvent. Not surprisingly, then, water is also a great producer of power. One day, "heavy water," or deuterium, may fuel the nuclear fusion process to provide us with great quantities of energy. Until then, however, we must use waterpower in simpler forms. Fortunately, there are many of those.

The continual circulation of water is called the hydrologic cycle. Unlike fossil fuels, water is not a one-time resource but is constantly recycled for a variety of uses from agriculture to power production. In 1975, hydropower plants were still producing three times as much electricity as nuclear plants. The installed capacity of presently operating hydroelectric plants in the United States is about 16 percent of the total electric power production. While the idea of using water to do work is certainly not new, its importance had been succinctly expressed by power expert Dr. Hans Thirring of the University of Vienna: "The use of water-wheels, though one of the earliest-known methods of power production, is still today the best of all methods as regards economy, efficiency, cleanliness, reliability, and inexhaustibility."

Like wind, water energy comes indirectly from the sun. It is the sun that creates weather, its heat causing the winds and vertical air currents. Air often contains water vapor, which condenses aloft to form rain or other precipitation. That which falls on higher ground obeys the law of gravity and begins its journey toward the lowlands, and ultimately back to the sea or other body of water it came from.

Only about one-third of one percent of the total solar energy reaching Earth is used in lifting water vapor into the atmosphere. Yet even this tiny fraction results in the continuous generation of about 600 billion horsepower! At present, there is no way known to tap the energy of precipitation falling through the air. Not until water is on the ground and flowing downhill can engineers use its power through hydroelectric or other power plants. Only about 3 percent is presently recoverable as useful power.

The Ancient Waterwheels

Water in motion possesses kinetic energy that can be converted to power with the proper machines. These were known as "mills" in early times, and many water machines operate equipment in which millstones grind grain—or something else. Thus waterpower began to take over these and other chores until then performed by man or animals.

Some historians estimate that it may have been 3,000 years ago that water mills began grinding grain in the Near East. Greece had waterwheels in common use long before the birth of Christ, and the poet Antipater exulted that "Demeter has ordered the water nymphs to perform the work of your hands." Interestingly, it was women whom Antipater told they had been freed from the drudgery of toil at the mill.

There were problems and delays, of course. Vespasian, the Roman emperor from A.D. 69 to 79, refused to have a water-powered machine built. He reasoned that the use of this natural resource would leave no work for the poor! Later Roman leaders were not so shortsighted. Overflow from the aqueduct of Trajan operated factories in the Janiculum, and corn mills were run by water from public baths and aqueducts.

The battles between hydropower proponents and environmentalists had their start in Roman times. In 395 Honorius and Arcadius issued edicts making it illegal to divert aqueduct water to run mills. But esthetics gave way to power production and Visigoth laws protected mills and their water supply, ordering anyone damaging them to restore the equipment within thirty days.

For centuries it was probably enough that a water mill operated at all. But over the years there arose a class of men known as "engineers," for the engines they tended. Many of these engineers were curious, dissatisfied, and never content until they could make their mills last longer, do more work, or be easier or cheaper to construct. The first mills had shafts at whose bottom blades were attached. The moving water rotated these blades, and gears at the top of the shaft in turn drove mill wheels. The name "Norse mills" has been given to these pioneering waterpower engines. While the primitive Greek mills had produced perhaps only ½ horsepower, the Roman engineer Vitruvius increased their power severalfold. A single Roman mill at Barbegal could grind 28 tons of flour in a ten-hour day, enough to feed 80,000 people.

Interestingly, it was the grain mill that led in time to the idea of a boat driven by paddle wheels. In A.D. 537, the besieged city of Rome mounted mills on ships in the Tiber River and ground enough grain to feed its people. From letting a river move a paddle wheel on a stationary boat came the reverse idea of having a powered wheel drive the boat.

England's great census, the Domesday Book of 1086, listed 5,624 corn mills in operation. Mills also powered saws, foundries, and ore crushers, and drained flooded mines. An interesting use of waterpower came in 1682 when French engineers built the largest waterwheel to that time. Designed to pump water for the gardens of Versailles, this complex installation produced about 75 horsepower. Perhaps the largest waterpower installation built until the days of giant hydroelectric plants was the huge overshot wheel built at Laxey, on the Isle of Man. Towering 70 feet into the air, the big waterwheel produced about 150 horsepower.

The first vertical waterwheels were known as "undershot" wheels, since they dipped the lower rim into the moving stream and were driven by the force of water on the paddles as they hit the bottom of

Drawing of the "Norse" mill. Great numbers of these primitive but useful devices were built.

their swing. Gradually it dawned on the engineers that feeding the water to the top of the wheel was a more efficient way of producing power, and men like England's George Smeaton made a true science of waterwheels.

An example of how proper design improved the output of water mills lies in Smeaton's measurements of undershot and overshot wheels. While the former delivered only 22 percent of the potential waterpower, the overshot wheel increased it to 63 percent. This tripling of output made waterpower an attractive project.

The Swedish engineer Christopher Polhem, who lived from 1661 to 1751, first applied waterpower to a factory in a scientific manner. As he pointed out, "There is great need of machines and appliances which will, in one way or another, diminish the amount or intensity of heavy manual work. This result can be most adequately achieved by the substitution of water power for hand work, with gains of 100 or even 1,000 percent in relative costs." Putting his belief to work, Polhem set up a metal-goods factory at Stiernsund in the year 1700.

This giant waterwheel was built at Laxey, Isle of Man, in 1854.

Its 100 artisans were aided by water-driven equipment in making plows, hammer heads, locks, clock wheels, tin plate, and pots and pans.

While we generally think of waterwheels as museum pieces centuries old, the mill at Stratford (England) built sometime before 1790 was still operating productively at the end of World War II. Only the postwar construction of a large reservoir made it necessary to dismantle the old mill, which was then re-erected on the grounds of a museum. The first important waterpower installations in America were made at Pawtucket, Rhode Island, in 1790; at Paterson, New Jersey, a year later; and at Fall River, Massachusetts, in 1813. These mills seldom used a head of water of more than 16 feet, and if more was available, the dam served an extra set of mills. (A head is the distance the water falls in producing power.) Such machines are called "low head." Present hydroelectric power plants in the United States have an average head of almost 100 feet. The highest head employed in a hydroplant today is that at Lac Fully, Switzerland, where the water drops more than 5,400 feet.

A new wave of scientific interest came on just before the middle of the nineteenth century, when Benoit Fourneyron in France and Lester Allen Pelton and James Francis in America greatly advanced the art of extracting power from moving water. The old wheel at Laxey became quaint when giant turbines produced thousands of horsepower to run cotton mills in New England.

The spectacular waterfall at Niagara was long eyed enviously by waterpower experts. Each second, 200,000 gallons of water plummet 200 feet to the bottom of the falls, and engineers estimated a potential of more than 6 million horsepower. The first harnessing of Niagara took place in 1875, when a canal was built to divert a small amount of water to drive water turbines. These produced mechanical power only, for the electric generator was still some two decades in the future. Nevertheless, this pioneering effort at Niagara produced 7,000 horsepower, transmitted by ropes and pulley to operate factories, a flour mill, and a brewery.

One thing remained to be developed and that was a method of carrying power produced by water mills an appreciable distance from the site. Crude attempts to do this with long shafts or rope-and-pulley arrangements were not successful; it took the development of electricity to give new life to waterpower.

In 1890 German engineers succeeded in transmitting electricity along wires from Frankfurt to Lauffen. Within a year, mining engineers in Colorado (using electrical generating equipment designed and built by George Westinghouse) installed a water turbine in the San Miguel River and sent the electricity it generated 200 feet up a mountainside to the Gold King mine near Ouray. The turbine generator, which skeptics claimed couldn't work, produced 100 horsepower of alternating current and triggered interest in similar development of Niagara Falls.

Thomas Edison, who had recently electrified New York City with his new direct-current power plant on Pearl Street, was asked to help out with the electric power plant planned for Niagara. Feuding with George Westinghouse, he argued strongly against the use of alternating current. But it was AC that won out, and in 1895 the Westinghouse 5,000-horsepower generators began producing electricity at

Niagara. A year later power was flowing through transmission lines and operating factories in Buffalo, New York. Waterpower was on its way.

By the turn of the century there was no stopping the exciting new hydroelectric plants. In 1907 a 10-megawatt hydroelectric plant was built at Big Bend, on California's Feather River. The Missouri was harnessed at Rainbow Falls, Montana, three years later by a 30-megawatt plant that sent electric power nearly 200 miles over steel suspension lines to Butte and Anaconda.

By 1913 Keokuk Dam had been built on the Mississippi, a river too big for engineers to ignore. Boasting more concrete in its structure than the volume of rock in all the pyramids of Egypt, this hydroplant produced 200 megawatts, even today a sizable power plant. But such giants were only a beginning.

During World War I construction started on the Wilson Dam at Muscle Shoals on the Tennessee River in Alabama. With a head of 134 feet and plenty of water, it seemed an ideal site and power was needed for the war effort. But when the war ended with Muscle Shoals unfinished, the project faded out for lack of funds. Not until 1932 and Franklin D. Roosevelt would it be given new life as the

Here is one of the many dams of the vast TVA project begun in the 1930s.
The Free Library of Philadelphia

huge Tennessee Valley Authority got underway. By 1944 a total of nine large dams and hydroelectric plants were extracting power from the mighty river system for a total of 624 miles.

"White coal," as hydropower became known, was being put to use out West too. From 1930 to 1936, during the worst of the Great Depression, gigantic Hoover Dam, or Boulder Dam, as it was known for some time, was built, creating Lake Mead and eventually producing about 1,500 megawatts of power.

The Columbia River, too, was tapped for hydroelectric power beginning in the 1930s. Bonneville Dam came first, but Grand Coulee was the crowning glory of all United States hydroplants, "completed" in 1941, but added to ever since. In 1975 additional turbines boosted its total output closer to its planned maximum of 9,200 megawatts. At the moment, however, Russia has the world's largest hydroelectric power plant at Krashoyarsk, generating more than 5,000 megawatts.

Waterpower at Work

Hydropower, now approaching a century of producing electricity for civilization, is not as glamorous as the newer nuclear power. Yet it produces about 4 percent of total energy needs in the United States, some three times as much as the nuclear contribution. U.S. hydroplants have a capacity of about 55,000 megawatts and produce 257 billion kilowatts per year. North America includes only about 13 percent of the world's potential waterpower but produces about 40 percent of the total power in the world. The three Pacific Coast states have about 34 percent of the total waterpower potential in this country, the mountain states another 24 percent. Thus 14 percent of our population has access to almost 60 percent of hydropower potential.

According to the Federal Power Commission, the total potential of hydroelectric capacity in the United States is about 179,000 megawatts, a bit more than three times that already developed. The total production of electricity would amount to more than 700 billion kilowatt-hours annually.

The federal government has constructed and operates more than half of the hydroelectric power plants in the United States. The U.S. Corps of Engineers has more than 17,000 megawatts in operation or under construction. The Bureau of Reclamation has a total of more than 12,000 megawatts, the Tennessee Valley Authority about 4,500 megawatts, and four other agencies operate lesser amounts of hydroelectric capacity.

Europe, with only 10 percent of the hydropower potential, also produces about 40 percent of the world total. Thus North America and Europe have 80 percent of the world's installed hydroelectric capacity, while the other 77 percent of the world potential accounts for only 20 percent of the total. Asia, with about one-third as much installed capacity as Europe, has developed less than 3 percent of its potential. Australia and New Zealand have developed about 2 percent, South America about one percent, and Africa so little that its present capacity could be increased 2,500 times!

It has been estimated that the world's hydropower resources are equal to the 1970 total world consumption of fossil-fuel energy. Potential capacity is about 3 million megawatts, of which Africa has

Grand Coulee Dam in the 1960s. Additions will one day make it the largest hydropower facility in the world. *Bureau of Reclamation*

780,000 megawatts and South America 577,000. Worldwide use of hydroelectric power totals less than 250,000 megawatts, roughly 8 percent of potential capacity. Should we be able to harness half of that, it could provide several times the total electrical needs of the world—unless increasing use outstrips the supply.

Obviously, there are many benefits to be gained from waterpower. No air or water pollution is caused by a hydroelectric power plant. There is no fuel cost, and no consumption of precious natural resources that are limited and needed for other uses. Waterpower is constantly recycled, driven by the huge solar power plant of evaporation, precipitation, and run-off. If hydropower is so great, then, why don't we make more use of it?

One easy answer is that environmentalists object strenuously to most proposals for new dams, notably Bridge Canyon and Marble Canyon in the Grand Canyon area. There are problems with hydropower, of course. Dams to provide the needed head of water pose environmental problems for wildlife and wilderness areas. A recent example is Egypt's experience with its Greater Aswan Dam, built over a long period and at tremendous cost. Although this power plant does produce great amounts of electricity, it has reportedly played havoc with the ecology of the area, depriving the land downstream of valuable silt and fertilizer and destroying fish production.

Filling of reservoirs with silt is a problem in long-term power production. It is also detrimental to irrigation, flood control, and navigation. Because the Columbia River carries little silt, Grand Coulee and Bonneville may provide power for more than 1,000 years. However, Hoover Dam lost 4.5 percent of capacity in thirteen years, and the Guernsey Reservoir lost 33 percent in twenty years.

Properly designed and carried out, hydroelectric power facilities can help rather than hurt the environment they are built in. Many such power plants have resulted in attractive recreational facilities, wildlife and fish enhancement, water supply, and water quality control. As an example of the multiple use of water resources, the Tennessee Valley Authority is a watershed area of about 41,000 square miles, including a population of close to 4 million people in seven states. Established in 1933, TVA has had great impact on its residents, providing not only electricity but 630 miles of navigable streams, recreation, water supply, and flood control.

Early in the development of steam engines, these fossil- or wood-fueled engines were used mainly to pump water to provide an artificial waterfall for waterwheels! Today, the reservoir of a hydroelectric plant can be used to provide cooling water for a neighboring nuclear power plant.

There is the other side of the coin as well. As a reservoir is "drawn down" to provide power, it may restrict its use for recreational purposes. Fishing and boating may suffer, for example. Downstream waterways are also affected by what happens at the hydroplant, and fisheries, recreational facilities, and farming all may be affected.

Another fact to be faced is the remoteness of many otherwise desirable hydroelectric sites, for it costs a great deal to build and maintain long transmission lines. There is also the reality that we have already made use of the most attractive hydroelectric power sites in our country. Indeed, even if we develop *all* available sites, the share that hydropower provides of our total needs must constantly shrink. This is because we are constantly using more fuel to take care of more people with more energy demands, and there is only so much waterpower available. Indeed, one projection for 1990 shows hydropower providing only 6 percent, rather than 16 percent, of electric power it now supplies. And that 16 percent is only half the contribution hydropower made in the 1930s. It is apparent that we must not only expand conventional hydroelectric use but also find new ways of working with nature.

The Pumped-Storage System

One of the problems of large steam-electric power plants is that they should be operated on a continuous basis for highest efficiency. The problem is that people don't use electricity on that basis, and there can be waste of unneeded power during periods of low demand, plus a shortage of power during peak demands. An electric plant cannot be turned on and off exactly with consumer demands, for the huge turbines take long periods of time to reach full speed and also to stop

Although hydroelectric plants change the environment, they can sometimes add recreational opportunities. *Bureau of Reclamation*

MONTEZUMA PUMPED STORAGE PROJECT
Arizona Power Authority

Schemes such as this near Phoenix, Arizona, provide a means for storing electric power for periods of peak demand. *Arizona Power Authority*

turning. What is needed is a kind of storage battery for excess power, and this is provided by a system called pumped storage.

Pumped storage was first used in Europe, as at the facility in Etzel, Switzerland. During off-peak hours, electricity is used to pump water back into a hydroelectric lake. The idea spread to the British Isles, and then to the United States, where there are now many pumped-storage facilities in operation.

There is a price to be paid for such storage of course, for there are friction and other losses involved in pumping the water up behind the dam for re-use. The price amounts to about one kilowatt-hour of energy lost for every three saved. However, the utility has saved two-thirds of an amount of electricity that would otherwise have been wasted. Pumped-storage facilities are being developed rapidly and in 1972 amounted to about 11,000 megawatts in operation or under construction. And an additional several hundred potential pumped-storage sites have been identified. By 1990 there will be a total of 70,000 megawatts of pumped-storage facilities, almost equal to the 82,000 megawatts of conventional hydroelectric capacity anticipated.

Pumped-storage projects built independently rather than in conjunction with a conventional hydroelectric plant pose special problems of an environmental nature. For example, a factor to be considered in pumped-storage siting is that of excessive leakage from the upper reservoir. Several upper reservoirs of pure pumped-storage projects have developed leaks in dikes, or in the structure of the earth or rock beneath the reservoir itself.

More Waterpower

Hans Thirring in 1956 nominated the Tsangpo River in eastern Tibet as the most attractive site in the world for hydropower production. The Tsangpo originates in Tibet's Lake Manosarovar, and flows due east between the Himalaya and Trans-Himalaya to eventually become the Brahmaputra River. Thirring believed that a dam built near Pe in Tibet would produce up to 333 billion kilowatt-hours a year. The magic in this number was that it was $3\frac{1}{2}$ times the electrical output in the United States the year Thirring made his proposal.

The problems of such a scheme are many. The site is far from any user of that much power, and the engineering of a tunnel and associated facilities would be difficult and expensive. But it is a challenging thought, to produce as much as 50 million horsepower at a single hydroelectric dam, ten times more than at the largest existing site. Perhaps there are ways of getting around the fact that only a limited number of natural sites are available as hydroelectric power producers. Engineers long ago suggested the creation of vast projects involving artificial hydropower sites. The forerunner was probably the Gibraltar project proposed by German engineer Hermann Sörgel in the 1920s.

Engineers had long been aware of a hydrologic phenomenon taking place at the Straits of Gibraltar. Since the Mediterranean Sea loses more water by evaporation than its tributary rivers (including the Nile, Po, Rhône, and Danube which drain into it), water is constantly rushing in from the Atlantic Ocean. This mighty "river" flowing into the Mediterranean is second only to the Amazon and represents tremendous amounts of energy. Unfortunately for the power engineer, there is no "head" at Gibraltar.

Sörgel's imaginative suggestion was to build a dam across the mouth of the Mediterranean from Gibraltar to Algeria! Shut off from its Atlantic Ocean water supply, the Mediterranean would slowly sink and, according to Sörgel's calculations, in a century there would be a head of 100 feet at Gibraltar. Power plants there, and at the mouths of all the tributaries, could yield a total of about 660 billion kilowatt-hours a year. In a pinch, the Gibraltar power plant could produce up to 500,000 megawatts of power since it would draw on the whole Atlantic Ocean!

For all its attractiveness, such a gigantic hydroelectric power system has not been built. It may never be built, for the dam at Gibraltar alone would require a volume of material 400 times that of Hoover Dam. Dropping the level of the Mediterranean 100 feet would also work some drastic changes on bordering countries. Transmitting power much farther than Spain and Morocco would be a problem, and those countries could hardly make use of so much electricity. Should the need for added land become acute in the Mediterranean region, however, the scheme might become feasible, for some 200,000 square miles would be added.

Egypt's Aswan High Dam, built with help from the USSR. *United Nations Food and Agriculture Organization*

Happily, there are more modest manmade hydropower schemes in the planning stages. In Egypt, whose Aswan Dam produces up to 2,100 megawatts of power depending on the vagaries of the Nile, a plan has been suggested to use a natural depression in the desert for the Mediterranean, and the Mediterranean rather than the Atlantic as a source of waterpower. This proposed plant would yield an estimated 4,000 megawatts (nearly twice the maximum output of Aswan) day in and day out, year in and year out, rather than fluctuating as does the output of the Nile.

The Qattara depression is in the Libyan Desert, an irregular-shaped sink about 180 miles long and 90 miles wide and at its lowest point more than 400 feet below sea level. The Qattara is about 50 miles from the Mediterranean. The proposal, first made in the 1930s, is to dig a channel (now probably with atomic explosives) from the Mediterranean to the northern edge of the depression. Water would then flow into the natural basin, driving turbines and producing electricity.

In time, of course, the depression would fill with water and the power production would stop—except for the high rate of evaporation in the desert. Detailed calculations indicate that when the water in the basin reached a level of 180 feet below sea level (at which time the surface area of the new lake would be 12,000 square kilometers, or about 4,000 square miles) evaporation would be taking place at the rate of 650 cubic meters of water a second. If that much flow was then admitted through the canal, 4,000 megawatts of power would be produced, a state of equilibrium would be maintained, and the lake level would remain constant.

The most recent proposals for the Qatarra depression project were made by Friedrich Bassler of Darmstadt Technical University in Germany. They include a nuclear dug canal (rather than earlier proposed tunnels), plus a pumped-storage facility to balance electrical output with load. Chemical industries are planned for the new lake, and the canal would permit ship navigation from the Mediterranean. As Bassler points out, the use of hydropower would permit export of large quantities of recently-discovered oil and natural gas in the Western Desert, thus giving Egypt a favorable balance of payments.

In 1971 scientists in Saudi Arabia proposed a similar hydroelectric scheme for the Dawhat Salwah depression near Bahrain Island in that country. The plan is to create an artificial lake separated only by dikes from the Arabian Gulf. Although the lake would be much smaller than the Qatarra depression, the absence of a long canal and its evaporation losses would make the Dawhat Salwah scheme more efficient. It would produce about one-sixth as much power as the Qatarra plant. There are similar topographic situations in many places in the world. For example, South American scientists have also urged development of hydroelectric projects utilizing such dry basins.

Do-It-Yourself Waterpower

While there is limited potential for more large commercial hydropower plants, many sites are suitable for much smaller waterpower plants for farms and homes. Because the idea of waterpower appeals to a growing number of environmentally-oriented Americans, a fair

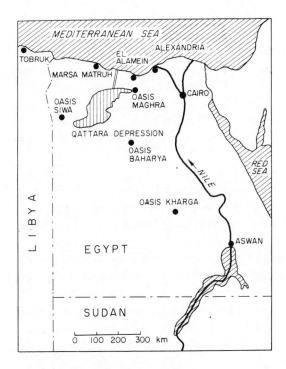

The proposed Qattara depression project which has even greater potential than the Aswan High Dam. *Solar Energy*

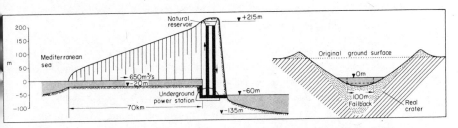

Cross section showing concept of the Qattara hydropower project. *Solar Energy*

number of hydropower plants are being constructed in rural areas, many of them by amateurs.

An old-fashioned waterwheel is better suited for pumping or other mechanical chores than for supplying electric power. Turbines are more efficient for electric power because their blades or vanes spin much faster, and thus don't require expensive gearing to run a generator. Of course such turbines are more complex and demand skill in their design and construction. However, once such systems are installed they offer free electricity for long periods with very little maintenance.

The amount of power in a particular site depends on the flow of water and the head. A tiny stream with little fall will produce minimal amounts of power. A stream that roars after a rain but tapers off in dry times produces power on a feast-or-famine basis and may not be a dependable enough source.

The basic waterpower equation is Total Horsepower = Flow Rate in cubic feet per second, times the head in feet, divided by 8.8. For example, 100 cubic feet per second, times a 10-foot head, gives 1,000. Dividing by 8.8 leaves 113.6 horsepower, an attractive amount. Of course no power plant is perfectly efficient, and after losses of friction and the like, something less than 80 percent of the theoretical power remains.

Attractive as do-it-yourself waterpower may seem, such a project should not be begun lightly, as the following warning, taken from a paper, "Small Water Power Sites" by Volunteers in Technical Assistance (VITA), indicates:

Flowing water tends to generate automatically a picture of easy, free energy in the eyes of someone who's looking for a source of "homemade" power. Don't be deceived. Harnessing water power is always going to cost *something*, and there are many factors that must be considered before you begin to dam up that babbling stream or rushing river running through your land.

As an example, the VITA paper says the following information is necessary before the neophyte can even contact a turbine supplier.

1. Minimum flow
2. Maximum flow

3. Head, or fall of water
4. Length of pipeline to get that head
5. Water condition (clear, muddy, sandy, acid)
6. Soil condition
7. Minimum tailwater elevation (channel leading downstream from dam)
8. Area of millpond above dam
9. Depth of pond
10. Distance from power plant to point of use of electricity
11. Distance from dam to power plant
12. Minimum air temperature
13. Maximum air temperature
14. Power needed
15. Map of site

Mechanical engineer Robin Saunders, writing in *Energy Primer,* estimates that the cheapest 10-kilowatt waterpower plant that could be put together would cost about $500. A fancier custom job might run as high as $10,000.

There are also environmental problems to be considered, even for a minimal project put together downstream of an idyllic millpond on the back-forty of a summer retreat.

While water or air pollution are not caused directly by a hydro-power plant, the construction of a dam does affect the ecology in the vicinity. Water backing up over land previously dry raises the water table there and makes it lower downstream. It drives away some wildlife and substitutes other marine life. As we have noted, silt is accumulated, and the area might also become a breeding ground for mosquitoes and other pests.

There are also legal considerations. Early water rights were ripar-ian rights, which allowed reasonable use of water to property owners whose land included or fronted on bodies of water. However, in the arid Southwest, water rights depend on appropriation. Here the first user has prior rights to the use of water for beneficial purposes on his land, whether or not his land adjoins the stream or other body of water. Compounding the problem is the fact that very few lawyers know anything about water law.

German colonists brought with them the knowledge of harnessing water power. This small mill is in Tovar Colony, Aragua, Venezuela.

Despite the fact that there are as many problems—and more—as there are attractions to home-built waterpower projects, a number of dedicated beginners have succeeded at the noble task and have been rewarded with a more-or-less dependable supply of electricity that is difficult to distinguish from the utility-bought variety.

Waterpower is a remarkable "alternative" source. It is remarkable for its long history of service, for the very low profile it maintains as a present supplier, and for the promise it offers for our energy future. Hans Thirring's words are as true as when he spoke them: for economy, efficiency, cleanliness, reliability, and inexhaustibility we must look long and hard to do better than the waterwheel.

4. Tidal Power

According to legend, Canute, an early king of Denmark and England, commanded the waves lapping the English shore to stop their ceaseless action. He failed of course, for the tides of the sea are as inevitable as the rising and setting of the moon, which plays a major part in the tidal phenomenon. In recent years the descendants of Canute have seriously considered generating electricity from the tides in the Severn Estuary, where the king once battled his enemies on the Island of Olney. This kind of harnessing of the tides could succeed.

As early as the eleventh century there were tidal mills operating in Great Britain, France, and Spain. One of these was in the Deben Estuary in Great Britain and was still in use after 800 years of operation. Such mills were used for mechanical power only, and they generated perhaps 30 to 100 horsepower. In 1682 a tidal waterwheel began pumping drinking water for London at Old London Bridge over the Thames. The sturdy waterwheel kept pumping until 1849, a span of 167 years. A few small tidal mills were also built in New England; one in Maine operated a sawmill. Hamburg, Germany, had a tidal power plant pumping sewage until about 1880. For the most part, such installations began to go out of existence in the late 1800s for economic reasons. They could not compete with other types of hydropower or fossil-fueled plants.

Tidal mills draw on a complex energy system involving the rotation of the Earth and the gravitational pull of sun and moon. There are tides even in solid rock, and the Earth bulges slightly in the direction of these attracting bodies, the tide moving across its surface like a ripple on the sea. If solid rock is attracted toward sun and moon, it follows that water will be pulled even more. And so we have the constant rise and fall of the tides that have affected life along the seashores, made poets ponder, and tantalized Canute and others with their obvious power.

Tidal action tends to slow the Earth's rotation. This slowdown is

Ancient French tidal mill at Saint Suliac. *Plenum Press*

no cause for immediate concern, for apparently the spin has slowed by only a few seconds per century since we began to check it. But even that tiny bit of friction constantly produces some 2 billion horsepower in the gigantic tidal engine.

Some energy from the tides is used up in scouring the ocean bottoms. More is wasted as heat, air movement, and noise. Only a tiny fraction of it is harnessed by turbines. If we were to capture it all, tidal power could handle something like half our present needs. We will never attain this highly desirable goal, however, for in just a few places do conditions permit practical and economic use of tidal power.

The Wave Machines

The turning of paddle wheels by the horizontal movement of a tidal stream has been exploited, and early tidal mills operated in this manner. Other suggested methods of extracting power from the tides include the lifting of weights and their subsequent fall as the tide goes out, and compressed-air schemes. While many test models have been made, and exaggerated claims and hopes continue, there are presently no successfully operating power-producing devices of this kind.

Sometimes confused with tidal power plants are wave engines, of which perhaps hundreds of designs have been invented and tested. These date back many years, and their popularity continues. For centuries clever minds have sought ways to tap the power suggested in the movement of waves across the ocean or large lakes. An early Prince of Monaco was interested in such a scheme and invested time and money in its research. Such engines have all the appeal of perpetual-motion machines and to date have been as unsuccessful. Yet current inventions are a wave-powered pump and a wave-powered propulsion system for ocean-going craft, both the brain-children of a Californian. The pump is a 60-foot length of plastic pipe anchored vertically in the sea. Wave movement can be made to pump water either upward or downward, and the inventor believes that a much larger pump, weighing about five tons, could produce as much as 50,000 horsepower in the region of the trade winds. Waves, of course, are produced by wind and not by tides.

The wave-propulsion engine consists of hinged flaps beneath a barge, or a string of barges, on the open sea. Wave motion causes the flaps to move up and down, driving the barge along at a speed the inventor hopes will be about four knots. The engine can be adjusted to drive the barge with the wind or against it. Just how useful such a machine would be remains to be seen.

Not to be outdone, an engineer at the Department of Mechanical Engineering at the University of Edinburgh, Scotland, has designed a different kind of wave machine. Built into an enormous craft perhaps 3,000 feet long, huge specially-shaped floats would bob up and down on the waves, operating a power cylinder about 30 to 60 feet in diameter. Not only would this produce power to drive the craft about, but it would also produce hydrogen from sea water. This hydrogen would be stored and brought to shore when the seagoing power plant was filled with the gas. The inventor claims that while previous wave machines extracted only 15 percent of the energy in the waves, his will boost this to an attractive 90 percent.

Despite the appeal of exotic wave machines, it seems most probable

Tidal power plant at the Rance Estuary, Saint-Malo, France. This is the only commercial tidal facility in the world. *Plenum Press*

Cook Inlet in Alaska is one of the potential tidal sites in the United States. *Plenum Press*

that further success in harnessing tidal power will come with tidal power plants like the two now in existence.

France's La Rance

The only sizable tidal installation in the world today is in France, on the estuary of the River Rance near Saint-Malo. Here a battery of specially-designed water turbines built into a great seawall produces about 350 megawatts of electric power. The turbines generate electric power in each direction, as the tidal flow reverses itself. The Rance project was completed in 1966 at a cost of about $365 per kilowatt of installed power. More expensive than conventional hydropower plants at that time, this nevertheless compared favorably with a fossil-fuel or nuclear plant, particularly when the free "fuel" in the form of water is considered. Each year the Rance plant produces more than 600 million kilowatt-hours of electricity for France's utility lines.

The Rance River project has been called a modest tidal plant by its engineers, and a much larger installation has been considered for historic Mont-Saint-Michel in Normandy, where tremendous tides isolate that great monastery on a vast beach daily. Using two dams running south and east of the island of Causey, this would produce between 10,000 and 15,000 megawatts of power and between 24 and 36 billion kilowatt-hours of electricity a year.

Russian Tidal Schemes

Russia's 400-kilowatt experimental plant at Kislaya Guba, 600 miles north of Murmansk, was begun late in 1968 and has been operating for several years. According to Russian engineer L. B. Bernstein, its design permits "matching the waves of tidal energy with the waves of energy consumption." Although less efficient than some tidal plants, it can operate at hours of peak demand for electrical power regardless of the phase of the tides.

Intersyzygial is the term given the difference in phase between the moon and man's solar time clock. Bernstein cites this as the reason the French gave up plans to build a much larger tidal plant than their

The size of this reversible turbine in the Rance Project is evident when compared with the engineer examining one of the huge blades. *French Cultural Services*

successful Rance River scheme in favor of a nuclear plant. The nuclear plant was simply much easier to regulate and match to periods of demand for power.

Bernstein points out that in most cases on a national basis it is not possible to balance the power output of tidal plants with need. He suggests, however, international cooperation in the form of joining the tidal energy of Cotentin Peninsula and the channels of England with the river energy of Sweden, Norway, and perhaps Russia as well. A similar approach could be taken by the United States and Canada using the Bay of Fundy, Cook Inlet in Alaska, and river hydroelectric plants of Alaska and the western continental United States.

Russia has a tidal power potential of 210 billion kilowatt-hours a year. The White Sea alone has a potential of about 40 billion kilowatt-hours a year, and this could be integrated into the power system of western Russia. Existing conventional hydropower plants would permit smoothing out the lunar fluctuations in tidal power output. Bern-

Kislaya tidal plant installed and in operation. This is a very small pilot project. *Plenum Press*

stein cites the high cost of tidal power-plant construction as the limiting factor in their development.

The solution used by Russian engineers was to construct the entire tidal power-plant building as a unit, then tow it like a barge to the site, and sink it into place. This approach copied the practice of building underwater tunnels this way and Russia's experience with floating reinforced concrete docks.

New Alaskan Bonanza

It has been said that almost as soon as Captain James Cook reported the huge tides of Cook Inlet in Alaska, engineers began to dream of harnessing the tremendous power in those tides. For most of the time the huge peninsula has been a remote and sparsely settled land with little use for electric power on a huge scale. However, that picture is changing, and at the present time there is interest in utilizing the tidal changes of Cook Inlet because of the growing population and energy needs of the new state. Even with plentiful petroleum in the region to fuel conventional power plants, the price that petroleum can bring in an energy-hungry United States may make a tidal plant for local electricity attractive 200 years after Cook made his big discovery. One proposed tidal scheme would involve 1,200 square miles of sea and produce an estimated 75 billion kilowatt-hours a year, about 7 percent of total United States electric power use in 1970.

Alaska's estimated yearly requirements are about one billion kilowatt-hours. A recent investigation suggested that partial development of the Cook Inlet potential could yield close to a billion kilowatt-hours a year plus a permanent road across Knik arm near Anchorage, plus deep water for Anchorage, reduced currents, and probably less ice hazard in winter.

Down Argentina Way

German engineers have studied the San José Gulf in Argentina as a potential site for a productive tidal plant. With an area of 780 square miles and tides as high as 30 feet, the site is estimated to produce about 12 billion kilowatt-hours a year. This would require damming more than four miles of the inlet and installing 400 turbines to convert the tidal rush into electric power. This would make it nearly four times the capacity of the Rance River project in France.

Canada's Bay of Fundy

The Bay of Fundy tidal installation was pioneered by a Canadian inventor about the time of World War I. By 1919 he proposed harnessing the tides on the Petitcodiac River and a decade later formed the Petitcodiac Tidal Power Company. This organization attempted to attract government development money but was totally unsuccessful.

At about the same time, Nova Scotians were investigating the possibilities of exploiting the Minas Basin tides. And also in 1919 the American engineer Dexter P. Cooper began his long work toward harnessing Passamaquoddy Bay, which straddles the Maine–New Brunswick border. Fifteen years later he became the most successful tidal-power proponent to that time, when President Franklin Roosevelt approved a $45 million project. However, in 1934 the time was still not right for tidal power. A year later, after the expenditure of $7 million on land acquisition and research studies, the project was halted amid charges that 'Quoddy was the greatest pork barrel yet.

Good ideas die hard, however, and in 1950 a joint U.S.–Canadian commission prepared another report on the Passamaquoddy scheme. In 1952 the scheme was reviewed by the U.S. Corps of Engineers, and in 1956 the International Joint Committee again got busy and spent three years in further studies. The final recommendation was for a 300-megawatt tidal plant using Cobscook Bay and Friar Roads as the tidal pools, and siting the power plant itself at Carryingplace Cove. In addition to the independent tidal plant, the committee looked at a combination scheme involving a conventional hydroelectric plant on the Upper St. John River in Maine, operating in conjunction with the tidal plant. This would give a combined total of 555 megawatts of power and generate more than 3 billion kilowatt-hours of electricity annually. All of this, of course, was at a time of relatively cheap power from other sources, and nothing came from the great visions of the U. S. and Canadian tidal pioneers.

In 1966 Canada drew up intergovernmental agreements creating the Atlantic Tidal Power Programming Board and the Atlantic Tidal Power Engineering and Management Committee. Then began a lengthy study of the Bay of Fundy Project. This effort lasted for $3\frac{1}{2}$ years and culminated in the International Conference on the Utilization of Tidal Power, held May 24–29, 1970 in Halifax, Nova Scotia. Canadian tidal experts were joined by experts from England, Russia, France, Germany, the United States, and the Netherlands.

Environmental assessments of tidal projects generally bring joy to all concerned. Of course tidal plants, by damming off waterways, could adversely affect fish life, as conventional dams can. However, it has been found that fish pass through the faster turbines in Scotland's hydroelectric plants; thus tidal plants may not have a bad effect on fisheries. Furthermore, clam and oyster cultivation are expected to benefit by tidal schemes because of reduced sedimentation and other changes. Interestingly, it has been proposed that Russia's Kislaya Guba tidal plant be used also as a herring trap!

One of the advantages of a tidal plant is its great reliability, even

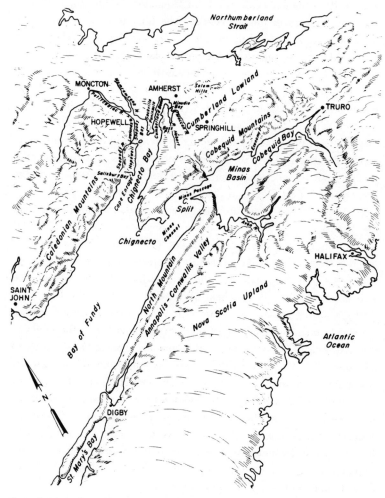

Canada's proposed Bay of Fundy tidal project. *Plenum Press*

higher than for a conventional hydropower plant. This is because the tides are very consistent and predictable, in contrast with the flood and drought sometimes associated with rivers tapped for hydropower. Reduction of tidal extremes might improve land use in the region, since flooding would be prevented. Tidal basins might also make for more lucrative fish farming. A little appreciated fact is that a tidal power plant affects the environment even less than a conventional hydropower plant, since there is no need to flood a large area of land to create the necessary head for power production.

Also on the plus side, tidal dikes might be used as roadways and thus shorten travel. Estimates of this effect in the Bay of Fundy project showed savings of as much as one million dollars yearly in the reduction of mileage. Forestry, for example, was shown to be a gainer. Cheap electric power in the area would lower some agricultural costs. And tourism might be promoted.

Although tidal power seems quite similar to conventional river hydroelectric power, and is sometimes called "salt-water hydro," the two sources are quite different. Indeed, one writer has said that the similarity between the two ends abruptly once it has been stated that in both cases the fluid motivating the prime movers is water in its liquid state.

Spring tides in Minas Basin run higher than 59 feet, and those in Chigneto Bay can be as high as 46 feet. There are two high waters and two low waters each lunar day of about twenty-four hours and fifty minutes. One of the obvious problems of tidal power plants is the irregularity of the tides as compared with the rising and setting of the sun. Tides depend on the moon, and the lunar day is almost an hour longer than the solar day which people gear their activities to. So the output of a tidal plant is constantly shifting in and out of phase with the demand for electric power. The answer, of course, is to integrate somehow the electric output of the tidal plant with that of a conventional plant to give a smooth and steady amount of electric power at all times. This may involve pumped-storage schemes, as are often used with conventional hydropower plants.

Canada's seismic zoning studies place the area of interest for the Bay of Fundy project in Zone 2, where earthquakes as strong as 5.7 on the Richter scale could be expected every sixty years or so. Another possible hazard could come from tidal waves, but this has not been allowed for in the design of tidal plant structures.

Experience in building dams and also in closing off inlets during wartime (such as Scapa Flow in Scotland during World War II) suggest that tidal dams can be built with no particular problems. However, because of the magnitude of the task of building a tidal plant, and the need for holding costs to a minimum to make the projects economical, very careful computer studies and simulations are required.

The simplest tidal power scheme is the "single-pool, one-way" scheme. In this, water fills the pool at high tide and is then allowed to run out through turbines and generate electricity. Thus electric power is produced only part of the time. Next in complexity is the "single-pool, two-way" scheme. Here water operates the turbine in each direction, as it flows into the basin or pool and as it flows back out. This requires a new or "reversible-flow bulb turbine," which was pioneered in the Rance River project in France. The two-pool scheme mounts turbines between these pools. One pool is filled by each high tide, the other is drained at each low tide. In this way water is available for power production on a continuous basis.

A total of twenty-three sites were examined in the Fundy project and the three best were selected: Shepody Bay, Cumberland Basin, and Minas Basin. Each of these in turn was evaluated as "single-effect," "single-basin, double-effect," and "linked-pair" schemes. A maximum installed capacity of 3,536 megawatts and an annual energy output of 7,560 million kilowatt-hours were estimated. A tidal plant site at the choicest location in the Bay of Fundy would save an estimated 10.1 million barrels of oil a year. However, the final findings were that at a 7 percent interest rate the Bay of Fundy tidal power plant could not compete financially with conventional power plants.

In his paper "Tidal Power in the Bay of Fundy," D. H. Waller, Assistant Director of the Atlantic Industrial Research Institute of Halifax, concludes with these remarks:

The ultimate decision about the future of tidal power will be a political decision. The Bay of Fundy report, for example, has provided the foundation—in the form of an engineering feasibility study, an economic evaluation, and an assessment of implications in comparison with alternatives—that can form the basis for the ultimate political decision about the development of the tidal power potential of the Bay of Fundy. That political decision will, of course, involve consideration of altered economic factors such as interest rates; but it will depend to a very great extent on the value judgments that are made by politicians. They must decide when the point has been reached at which society is prepared to accept the additional costs of tidal power in order to avoid the effects of pollution due to alternative power sources.

Although that time has not yet arrived, there is little doubt that the increasing demands of an exploding population on a finite, fragile, and already abused environment will ultimately call into use any alternative that promises that the quality of man's environment does not have to be sacrificed as the price of technological advancement.

Like geothermal and hydropower, tidal power offers at least a modest potential for development beyond its present scope. The tides can never take over a major share of our need for electricity, but in certain geographic locations they can contribute appreciably. Tidal power is clean, and attractive in many other ways. At the moment, as Waller points out, it is not yet economically attractive. Despite the ancient adage, however, the tides do wait for man.

5. Sea Thermal Energy

There are many ways to get power from the sea, and we have looked at only a few of them. Other methods include tapping the major currents to drive turbines, exploiting the ocean's salinity, and harnessing "biopower" produced by living marine creatures. However, the best estimates for these ways suggest only a modest output of energy. What is needed for effective waterpower is an energy source of great magnitude. That source seems to be "sea thermal energy," solar heat stored in water.

As with many "revolutionary" ideas, sea thermal energy turns out to be quite old. French scientist Jacques d'Arsonval in 1881 wrote about it in great detail, suggesting that we would someday mine the seas, rather than Earth itself, for energy. His idea was basically an engine operating on ammonia vaporized by the heat of surface ocean water and condensed by cold water brought up from the depths.

As many know from swimming in deep water, water at a depth is cooler than that on the surface. In the oceans this phenomenon is carried to extremes. Water cooled to very low temperatures in the polar regions is denser than warm water; it sinks to the bottom and tends to stay there. There is little mixing of cold and warm except very near the surface, so there is a constant reservoir of chilled water lying under the warm tropical and subtropical oceans and seas.

The STE Pioneers

Not until the late 1920s did a countryman of d'Arsonval put the sea thermal energy idea to a test. Georges Claude, noted for discovering and exploiting neon gas and acetylene welding, constructed several sizable sea thermal energy plants and eventually succeeded in producing a modest amount of power from them. He did this at the expense of his own personal fortune, plus the money of backers, and a great investment in time and effort on his part. None of his sea thermal energy schemes was economically successful, and he died a failure as an exploiter of sea energy.

Claude noted in his writings that he learned of this prior work only after he himself was deeply involved in the challenges of harnessing sea energy. Had he known so many others had explored this path he probably would not have been interested, he said, preferring to take on tasks others had not yet done groundwork for. Of all his scientific research and engineering projects, surely this was the hardest he ever tackled.

Claude had trouble convincing people that water could boil at temperatures lower than 212 degrees F., the boiling temperature at atmospheric pressure. If you climb a high mountain you will find that it is difficult to cook food. This is because at the lowered air pressure at altitude, the water boils away at a lower temperature, and thus does not get the food as hot. Claude delighted in pointing out that if one produced a sufficient degree of vacuum he could boil ice water. How odd, he noted, that one could fall into boiling water and catch cold! In Belgium, Claude operated a steam turbine at 5,000 revolutions per minute and produced about 60 kilowatts of power— with a temperature difference between steam and condensing water of only 20 degrees C. This took place in the early 1920s.

Dr. Donald Othmer, of Brooklyn Polytechnic Institute, is among those involved in sea thermal energy projects today. He differs from his colleagues in that he was at the University of Michigan as a young graduate assistant in 1925 when Georges Claude gave a laboratory demonstration of his proposed sea thermal plant:

Claude's equipment included a small tank for warm water and one for cold, a vessel to which the warm water was admitted to undergo flash evaporation, a small steam turbine driven by the low-pressure steam that was forming, a condenser having the cold water in direct contact with the turbine exhaust steam, and a vacuum pump for air removal. A small generator was driven by the turbine and was wired to a small electric bulb, which lit as the house lights went off—and the audience cheered!

By 1927 Claude tackled the ultimate challenge, the one that would at last defeat him in spite of his noble effort. Off the city of Matanzas, Cuba, he succeeded in sinking into the sea a specially-made cold-

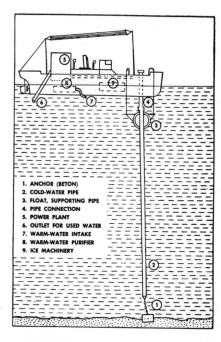

1. ANCHOR (BETON)
2. COLD-WATER PIPE
3. FLOAT, SUPPORTING PIPE
4. PIPE CONNECTION
5. POWER PLANT
6. OUTLET FOR USED WATER
7. WARM-WATER INTAKE
8. WARM-WATER PURIFIER
9. ICE MACHINERY

Diagram of Claude's experimental sea thermal energy plant aboard the ship *Tunisie.*

water pipe 6 feet in diameter and $1\frac{1}{4}$ *miles* long. Success came only after two failures, in each of which a fortune in pipe disappeared into the sea and had to be replaced. But the third time was the charm and Claude was able to pump up cold water from the depths reached by the outer end of his pipe. With a small temperature difference between warm and cold water, he produced about 22 kilowatts of power with his turbine.

In December of 1930, Claude addressed the American Society of Mechanical Engineers (who had recently presented him with its Fiftieth Anniversary Medal for his work prior to his sea thermal energy experiments). He described his trials and final success at Matanzas, and then told the audience of his planned 25-megawatt plant at Santiago, Cuba. He concluded his remarks by saying, "Humanity has from now on the certainty that its industries will never lack the precious energy that actuates them." He was not successful in building the proposed plant at Santiago, however, and even a modest floating facility on the ship *Tunisie* was a failure too.

Claude was a half-century before his time. He had produced only a token amount of power with his abortive Cuban plant. A French scheme off West Africa in the 1950s was designed to produce 7 megawatts. It was never successfully operated, and the sea demolished the huge undersea pipes feeding the condensers just as it had wrecked those of Claude.

Another Chance for the Sea

For a long time sea thermal energy was considered nonsense, in the same class with antigravity and perpetual motion machines. Then in 1961 Asa Snyder, a vice president of research at Pratt and Whitney Company, proposed the same scheme that had broken Claude's spirit and his bankroll.

Snyder reviewed the work of Claude and his successors, then presented his reading of the situation. Using a 30-degree F. temperature difference between hot and cold sea water as the minimum needed to be useful, he defined a "thermal-difference belt" between latitudes twenty degrees north and twenty degrees south. Each square foot of this huge expanse of sea, according to Snyder, receives one kilowatt-hour of energy a day (enough to operate ten 100-watt light bulbs for an hour). Estimating electricity costs at two cents per kilowatt-hour, he arrived at a value of $4 per square foot per year. Multiplying this

Section of cold water intake pipe of French experimental sea thermal energy plant off Abijan, Africa. *Energie de Mers*

amount by the total number of square feet, he arrived at the total value of solar energy received: $52.7 quadrillion! After realistic discounts for weather, inefficiencies, and so on, there still remained about $10.5 quadrillion worth of energy in the tropical and semitropical waters of the world.

Snyder estimated only a 3.5 percent overall efficiency for the sea thermal energy engines he had in mind, yet even this drastic reduction left some $370 trillion worth of energy. At 1961 figures, that was 92 trillion kilowatt-hours, equivalent to about 131 billion tons of coal a year. For comparison, in 1961 the world consumed less than 3 billion tons. Should we apply all the energy available from the sea to desalting water, we could produce more than 920 trillion gallons of fresh water a year.

Snyder suggested the sea thermal energy plant mainly for the underdeveloped areas of the world. He named the places that could exploit this energy: Central America, the Caribbean islands, the north and central coasts of South America, Africa from Mauretania to the Congo and from Somaliland to Madagascar, the south coast of Arabia, Ceylon, Indonesia, New Guinea, Borneo, the Philippines, Taiwan, and Australia. Most of these places were short of fuels and fresh water was an increasing problem, but sea thermal energy offered these countries self-sufficiency.

In a land still buying all the gasoline it needed for thirty-five cents a gallon or so, and with natural gas a cheap and clean source of heat and power, Snyder's grandiose plan had all the impact of a sigh in a windstorm. And the underdeveloped lands he thought could benefit had no money to conduct the needed developmental work. But there were a few other dreamers who had caught sea thermal energy fever along with Snyder.

Design for a Sea Power Engine

At about the time Snyder was publishing his ideas, J. Hilbert Anderson and his son James, Jr. had also begun research toward a sea thermal energy plant. Already involved in the design and construction of geothermal power plants, they were convinced that capital was the only obstacle to a successful sea power plant.

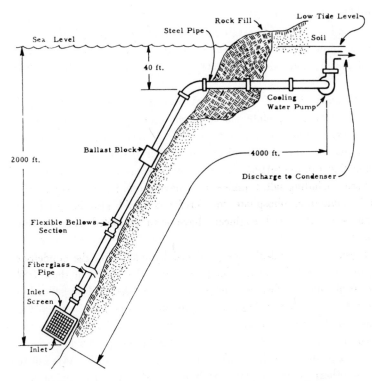

Snyder's STE concept, installed on a steep shoreline. *Solar Energy*

They maintained Claude had failed because he used water as the working fluid for his engine. D'Arsonval had suggested ammonia, which would be more efficient at the very low temperature differences provided by sea water. The Andersons proposed to use propane in their turbine, permitting reduction of the turbine diameter from 30 feet to 6 feet, cutting costs and making construction simpler.

Using existing fuel costs, the Andersons determined that a sea thermal energy plant built for $580 per kilowatt would be competitive with conventional power plants. Next, they calculated the costs for their design and arrived at an encouraging $168 per kilowatt. They also noted that the potential for power from the sea was about 200 times the projected world need for the year 2000.

Their 100-megawatt plant would be a floating barge, anchored in a suitable spot offshore. About 300 feet below the surface, large, ocean-warmed boilers would heat propane to drive the turbines and produce electric power. Above the boilers, equally sizable condensers would cool the propane with cold water piped up from a depth of 2,000 feet.

Obviously the STE plant would consume no irreplaceable fuels. Neither would it pollute the atmosphere or the ocean. There were other benefits not as readily obvious. One was the possible recovery of valuable minerals, chemicals, and other materials from the sea water that ran the turbines. Some substances are already mined from the sea, including salt, bromine, and magnesium. Potash, magnesium, and ammonium phosphate for fertilizers could also be produced from sea water, and engineers have even extracted gold in tiny amounts.

Fresh water could also be produced by modifying the condenser section of the STE plant. The Andersons calculated that their 100-megawatt pilot unit would produce as much as 800 million gallons of fresh water per day, using one-half the power produced by the plant. Costs of this fresh water, piped forty miles to shore, would be less than four cents per 1,000 gallons for 800 million gallons a day.

Other prospects included the siting of industries on or adjacent to the sea thermal energy plant. Aluminum plants would be prime candidates because large amounts of electric power are needed in the refining process. Sea thermal plants could be sited close to supplies of bauxite ore and save appreciable amounts in shipping costs. Rather than ship bauxite all the way to Canada where it is refined into aluminum, using that country's cheap hydroelectric power, the Andersons suggest building a sea thermal energy plant off Jamaica, in the British West Indies. The same could be done with bauxite deposits on the northeast tip of Australia, and in Ghana and the Guianas. Also noted was the fact that oxygen is a cheap byproduct of water desalting.

There would be no need for expensive real estate, or any real estate at all, for that matter. No longer would it be necessary to browbeat or bribe certain regions to host undesirable power plants. Offshore STE plants would not even be visible to those on the main-

Model of the floating Anderson sea thermal energy plant. *J. Hilbert Anderson*

land using the power output. The biggest surprise in the list of bonus byproducts the Andersons read off was that of enhanced commercial fishing grounds. The sparkling clear waters of the tropics and sub-tropics got that way because in the upper 300 feet or so of the oceans the nitrates, phosphates, silicates, and other elements necessary for life processes have largely been taken up by fish and other living things. Deep down, however, there is an abundance of these nutrients.

The greatest fisheries in the world are found where natural ocean upwellings bring up nutrients from the deep to attract fish and other marine creatures. The sea thermal energy plant produces its own miniature upwelling, thus the water surrounding the plant might well become profitable fishing grounds for commercial fleets. Nutrient-rich sea water would be confined to a certain area, and be effective in keeping the fish in one spot.

There were other interesting proposals by the Andersons. Rather than being useful only to underdeveloped nations, sea thermal energy plants could be built close to American consumers. Georges Claude had used Cuba as the site of his land-based experiments; the Andersons suggested Puerto Rico as an ideal spot, with deep water within five or ten miles of shore, and temperature differences of about 40 degrees F. Puerto Rico also has a rapidly expanding population and needs a large power supply. Another proposed site was twenty miles offshore from Miami, Florida. There the surface water averages 79 degrees F., and at a depth of 1,312 feet the water is a chilly 43 degrees F.

The Andersons reported that there are at least 18,000 miles of tropical coastline worldwide. Even off highly developed areas such as America's Atlantic seaboard, the Gulf Stream would provide power. Texas could get electricity from plants in the Gulf of Mexico from Tampico south, California might draw on plants off the tip of Baja California, and the Mediterranean off Israel would provide power and fresh water for that region. There are eleven countries adjacent to the Indian Ocean, fifteen adjacent to the Pacific, and twenty-nine around the Atlantic. Ten of these fifty-five countries are only one mile from a suitable sea thermal energy source! Another eleven are within ten miles, and ten more within fifteen miles. Only a half dozen are seventy or more miles distant. Thus STE plants could have a global impact on both developed and underdeveloped lands.

After a decade of planning and publicizing their STE plant, however, the Andersons had little more to show for their effort than their own stronger conviction that they were on the right track. In 1971 Alaska's Senator Mike Gravel, a proponent of alternative energy sources, inserted reports of their work in *The Congressional Record*. And other people were finally beginning to be aware of the sea's great promise.

Science Looks to the Sea

For about a decade, the Andersons labored at their designs, calculations, and model building almost completely unnoticed. In part this was because the Andersons are engineers and not skilled entrepreneurs. Furthermore, few in high places took seriously the idea of running a steam engine in the sea. But concerns about energy shortages finally got sea thermal energy a fair hearing from people in a position to hasten its development.

In April of 1971 Dr. William Heronemus and others at the University of Massachusetts made a proposal to the National Science Foundation for a study of pollution-free energy systems. Along with wind energy the group proposed what it called the "ocean thermal difference process" as a potential candidate for serious development. In doing its homework, the group retraced the work of d'Arsonval, Claude, the French after World War II, and particularly the research and development of the Andersons. Impressed, the NSF in 1972 granted the Massachusetts group $340,000 under the Research Applied to National Needs (RANN) program.

Thinking even bigger than the Andersons, Heronemus proposed to anchor a huge submerged facility offshore from Miami, where Gulf Stream heat would drive a dozen turbines mounted in two streamlined hulls. According to its designers, this STE plant would produce about 400 megawatts of electric power. This "Mark I" effort would be no toy, Heronemus pointed out, comparing it to the 185-megawatt nuclear plant at Rowe, Massachusetts, and the Vermont Yankee 534-megawatt plant. Furthermore, the STE plant could also use its power to distill sea water and electrolytically separate it into hydrogen and oxygen. These gaseous fuels would then be piped ashore in hoses for use as needed.

Heronemus and his colleagues found that there are 560 miles of Gulf Stream (from the tip of Florida to a point east of Charleston, South Carolina), wide enough to provide about 8,000 square miles of ocean with temperature differences of 30 degrees F. all year. This area could site four thousand 400-megawatt Mark I STE plants that would match the power output of 4,500 1982-class light-water nuclear reactors. For comparison, less than sixty nuclear plants are operating in the United States today. STE plants in the Gulf Stream alone could provide all the power needed by our country.

Recognition for the Andersons came in the form of a contract to design, build, and test a small working model of their sea-power plant. Results indicated that it should operate on temperature differences of as little as 3 degrees F. Of course, the greater the difference in surface and deep-water temperature the more power the turbine would produce.

Questions and Answers

In July 1973 the NSF also funded research by Dr. Clarence Zener and two colleagues at the Carnegie-Mellon University in Pittsburgh. Zener is noted for his invention of the Zener diode, a solid-state device much used in electronics. According to physicist Zener, it would be possible to harvest ocean solar energy and produce electric power at a cost competitive with fossil fuel plants. Solar sea power could provide an appreciable amount of the electric power needed by the U.S. and other countries.

A strong point in sea thermal energy's favor, especially when it is compared with other solar energy schemes, is its twenty-four-hour capability. It does not depend on the sun shining, simply on solar heat stored over time in the sea.

Critics have also pointed out that enormous quantities of water must pass through the STE boiler to produce electricity. To this, Zener replied that so do enormous quantities of water pass through hydroelectric power plants, perhaps the finest power producers in operation. As a matter of fact, an STE plant produces as much power from a given volume of water as that water would produce falling about 100 feet at a hydroelectric dam!

An Idea Whose Time Has Come?

In spite of the fact that there are as yet no operating STE plants, the technology is not the pioneering venture it might seem. Oil companies have for years built huge marine oil-drilling rigs that make a good starting point. In his work for the NSF, Dr. Heronemus involved the United Aircraft Corporation in design work on the turbines needed for STE plants. And Lockheed Missiles and Space Company has proposed a pilot STE plant weighing 300,000 tons and producing 160 megawatts of power at a cost of 2.7 cents per kilowatt-hour, competitive with fossil-fuel plant costs.

Lockheed pointed out what STE engineers had known for decades:

The oceans represent a virtually inexhaustible reservoir of heat energy, replenished continually by the sun. This reservoir can be tapped twenty-four hours a day and not just when the sun shines or when the wind blows. While giving 3 percent as the expected efficiency of its plant, Lockheed pointed out that the other 97 percent of heat energy is not wasted, as is the case with fossil fuels and other engines. Instead, it is returned to the sea to be used again at another time.

The design produced by Lockheed consists of four power modules attached to a central platform. A cold-water intake pipe 1,000 feet long and mooring attachments capable of anchoring in 20,000 feet of water were provided. Electricity would be transmitted to shore by cable in the form of direct current, to a maximum of eighty miles. As alternatives to electricity, the OTEC plant would have the capability of producing gaseous or liquid hydrogen, metallic hydrides, ammonia, or other forms of stored energy that could be transported ashore for use as needed. Echoing the earlier suggestions of the Andersons, Lockheed suggested bringing industrial plants to the floating power plant, including metal refineries, fertilizer operations, and synthetic fuel plants.

Lockheed's gigantic power plant would be made of slip-formed concrete, reinforced for additional strength. This material was chosen because of the long history of its use in offshore structures. For an idea of the size of the structure, the high-rise portion above water is 60 feet in diameter. Water would be moved through the power modules at the rate of 28,000 cubic feet per second, a Niagara in each module. And if the entire project boggles the mind, think for a moment of the size of a Hoover Dam.

Because it would be a seagoing "vessel," the Lockheed plant was designed to comply with the rules and regulations of the U.S. Coast Guard with regard to navigation and crew safety. Lights, foghorns, radar equipment, and communication equipment would be provided, as well as watertight doors, alarm systems, emergency breathing gear, and other safety features. The crew of thirty-five would be operations and maintenance specialists with a level of skills comparable to personnel working in a fossil-fueled plant.

Costs of the Lockheed OTEC power plant were considerably higher per kilowatt of capacity than those predicted by Anderson and others. For instance, using titanium in the heat exchangers, the estimate was $2,594 per kilowatt, several times that of a nuclear plant. However, electricity would still be produced at a cost of about 2.3 cents per kilowatt hour and this would be competitive with oil as priced in 1975.

By using aluminum rather than titanium, first cost could be cut to only $1,350 per kilowatt, just over half that of titanium. Other design changes, plus operation of the plant in equatorial water where there is a higher temperature difference, would reduce per-kilowatt costs to perhaps as low as $350, half that of a nuclear plant. This would be a mixed blessing, however, since it is a long way from the Equator to a consumer of electricity in the United States.

Fossil and nuclear plants convert raw materials into heat, most of which is dumped into the environment to create thermal pollution of air or water. Conventional plants also produce waste products with varying degrees of undesirability. But the OTEC plant adds no heat to air or water and does not contribute any pollutants, except for the noise of its operation. The noise should be no worse than that of a conventional plant, and since it will be well offshore it would not be as much of a nuisance. The only possible contaminant would be the ammonia working fluid. In case of a leak, this would be diluted with the ocean water and biologically taken up with no known adverse effects. Concrete, steel, and titanium were chosen because of their minimal corrosion.

As heat is removed from the ocean to operate the OTEC plant, more incoming solar energy will be absorbed by the ocean. Over the ages, a balance has been set by the environment, and there is no danger that the seas would be cooled down. In fact, some scientists think there might be a slight *increase* in ocean temperature, but not on a scale to cause any environmental changes.

The Doubters

Just how fast, and how far, the STE program will go remains to be seen. Admittedly such projects are very complex and as yet almost completely untested. Environmental impact is only one area that

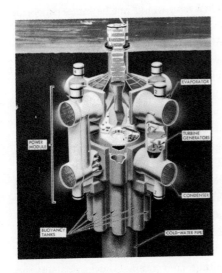

Artist's drawing of Lockheed's OTEC plant. The scale of such a project matches that of a large hydropower dam. *Lockheed Missiles and Space Company*

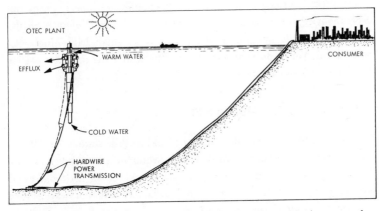

Sunshine, ocean water, and human ingenuity provide electric power for coastal areas. *Lockheed Missiles and Space Company*

could pose problems. The Andersons and Clarence Zener feel that *more* solar heat would be retained by the ocean as cold water was pumped up from the depths. Thus man would not be tampering with the gigantic ocean system and drawing away stored ocean heat as he has drawn or overdrawn on his forests and fossil fuels. However, for all the sea thermal energy's seeming attractiveness as a clean and safe power plant, there are those who worry that such a plant may upset the ocean's equilibrium and produce intolerable changes in the environment, such as a new ice age.

Others damn sea thermal energy with the faintest praise. In fact, a study done for the Senate Commerce Committee by Robert R. Nathan Associates in 1974 almost totally dismissed any possibilities for STE. In a report of over 100 pages, entitled "The Economic Value of Ocean Resources to the United States," the consultants devote only two short paragraphs to sea thermal energy, pointing out that Claude's experimental efforts were unsuccessful and that suitable sites are available only in some parts of the Gulf of Mexico. Instead they see far greater potential in siting conventional nuclear plants on floating ocean stations, or using deuterium from sea water to fuel fusion power plants, although the report admits that fusion is not expected to be available until well after the turn of the century.

Happily, there is an approach sizable enough to properly check out STE theories, yet modest enough not to cost several fortunes in the process. Even its name carries the feeling of small beginnings: the solar pond. Yet it too is an idea tracing back into the past, far earlier than the work of Claude.

The Solar Pond, A Modest STE

The idea of the solar pond is almost as old as this century. In 1909 a Pennsylvania engineer named Frank Shuman proposed a 1,000-horsepower solar engine run by water heated in an artificial pond covering four acres. No starry-eyed dreamer, Shuman was a very practical designer and he worked out his concept in a brilliant manner.

The pond site would first be leveled and rolled flat with heavy equipment. It would then be coated with black asphalt, serving the double purpose of waterproofing the soil and making it a good heat absorber. A structure resembling a very low hothouse would support panes of window glass about a foot above the soil. Three inches of water would then be run into the entire pond and a thin layer of melted paraffin would flow on top of it. This waxy layer, plus the glass above, would help retain solar heat inside the collector.

Shuman estimated that the sun would warm the water to just over 200 degrees F. Because the sun would be unavailable part of the time, he provided a huge storage tank 30 feet high and 75 feet in diameter to supply his turbine when hot water was not available directly from the pond itself. Because of a partial vacuum in the system, the 200-degree water would produce steam to drive the specially-designed turbine. Waste water would of course go back into the solar pond for reheating.

The turbine would operate on a temperature difference of about 100 degrees F., far more than that available to the proposed STE plants, and thus would be efficient. With his 160,000 square feet of collector to provide solar heat, Shuman calculated an efficiency of between 6 and 7 percent. Since the sun would be shining only about one-fourth the time, however, the overall efficiency would be only about 1.5 percent.

Estimated cost for the solar collector was about twenty-five cents a square foot, or about forty dollars per horsepower of installed cost. In 1909 that was a high enough price that Shuman apologized for it, explaining however that it was justified because fuel for his turbine would be free.

After a few small-scale tests, Shuman realized that the technology available to him was not up to the challenging task of the solar pond. He changed his approach to a concentrating-type collector to provide higher temperatures and by 1913 had produced a remarkable solar irrigation pump that developed about 60 horsepower on the outskirts of Cairo, Egypt. But he had blueprinted the solar pond and associated turbine. Those who reinvent it today are drawing on the bold concept of Frank Shuman, another solar engineer a half-century and more ahead of his time.

For almost forty years the solar pond idea seemed to have evaporated. Then in 1948 Israel's Dr. Rudolph Bloch suggested a unique concept that came to be known as the "reverse thermal gradient" solar pond. As the name implies, this is the opposite of the ocean situation in which water is cold at lower levels and warm on top. This arrangement works against the solar pond engineer because the warm surface water tends to lose its heat rapidly. Bloch's reverse thermal gradient scheme would result in the water at depths being hotter than that on the surface!

The problem seems as tough as trying to balance a pyramid on its point, but by a clever trick he succeeded. The germ of the idea came from a lake in Hungary where the reverse gradient had been produced naturally. Because of salt concentration in the lake, increasing with depth, movement of heated water upward was suppressed. Ten years after Bloch got the idea, Israel's National Physical Laboratory under Dr. Harry Tabor began to study the solar pond with reverse density gradient. Within a year, temperatures of 60 degrees C. had

Experimental reverse gradient solar pond in Israel. *The National Physical Laboratory of Israel*

been attained in an old evaporation pond. A year later a larger pond of about 625 square meters on the southern shores of the Dead Sea was carefully treated with magnesium chloride and achieved a bottom water temperature of 90 degrees C. with the surface water at air temperature. And 90 degrees C. is 194 degrees F., approaching the boiling point.

A reverse gradient pond is a difficult thing to produce, and even more difficult to maintain. However, with some care the salt concentrations can be maintained for a few weeks. Tabor theorized that if hot water could be extracted from the bottom—without upsetting the density gradient—and used to run an engine, a pond of one square kilometer would produce about 40 million kilowatt-hours of electricity a year. This would call for an overall efficiency of about 2 percent, slightly less than predictions for a large STE plant. However, the one-kilometer pond would drive a 5-megawatt power plant, enough to serve about 5,000 people with electricity. If the pond was used simply for heat to provide house heating and cooling, and domestic water, it would save about 50,000 tons of fuel oil a year.

Professor Julio Hirschmann of Valparaiso, Chile, later designed a solar pond (minus reverse gradient) which he estimated would produce 50 kilowatts of electricity, plus 4,500 gallons of distilled water a day. However, neither Chile nor Israel has reported further on the solar pond as a power plant.

Tabor suggested that the pond would also be useful as a producer of salt, since it would be more efficient than the conventional evaporating beds normally used. Even this approach has not yet been tried, but the solar pond as a producer of plain hot water is being tested at Grants, New Mexico.

ERDA's Lawrence Livermore Laboratory (formerly operated by the AEC) and the Sohio Petroleum Company of Cleveland, Ohio, teamed in the design and construction of the first sizable solar pond. An estimated six acres of plastic-covered shallow pond would provide the necessary 500 gallons of 140-degree F. water hourly in the uranium ore milling process.

The basic idea parallels that of Shuman's 1909 design, but the materials are different. Instead of an asphalt bottom, the Grants pond uses a black heat-absorbing material laid on the leveled soil. Rather than covering all the available space, the pond would be arranged in rows of basins about 12 feet wide and 200 yards long, with space between for service work. A single layer of plastic would prevent evaporation of the water, and above that two or three more layers of plastic would provide a greenhouse effect such as Shuman had been planning.

The purpose of the New Mexico solar pond is to provide industrial hot water but it may also be used to test the feasibility of the age-old dream of producing electricity from a solar pond. In this application, the hot water would heat a secondary medium such as Freon, which in turn would operate a specially designed turbine to generate electricity. While it may not seem poetic to use solar energy in a process to make nuclear fuel, there will be an appreciable saving in oil formerly used to heat the process water. And since industrial heat amounts to perhaps 20 percent of our total energy needs in the United States, the solar pond may well fill a great need.

Whether solar energy will be a big frog in a little pond or a 100-megawatt plant almost lost in a vast ocean remains to be seen. But unless a number of obviously capable scientists and engineers are completely wrong, the sea that has been man's avenue of exploration and travel, and a plentiful source of food, may before long supply him with great amounts of energy that may well be inexhaustible. The next few years should tell.

PART THREE

Energy from the Wind and Sun

6. Power from the Wind

Wind is an inescapable fact of life, a phenomenon that everyone from sailors to girl watchers depends upon. The fact that wind is solar energy once removed may have escaped our attention, and wind as indirect sun power takes some getting used to. A moment's thought will clear up any doubts, however. It is the sun that heats the Earth, and the Earth in turn heats the air. Sunshine imparts but little heat to the atmosphere as it passes through, but warm land or water do heat the overlying air by conduction.

Because of varying terrain, and because of clouds, the Earth is heated unevenly. This in turn causes the movement of air as Nature tries to equalize pressures. She never succeeds, but a great deal of huffing and puffing by the north wind and other breezes takes place in the process.

Wind energy is a form of kinetic energy, the energy of moving molecules of gas that cause a reaction in whatever they strike. A windmill functions basically the same way as a water mill; the difference is in degree. Water is quite dense; air is not very dense at all, and we are seldom aware of it until we consider it as we are doing now. Air is only about $\frac{1}{800}$ as dense as water, and there is not nearly as much kinetic energy in a cubic foot of moving air as there is in the same amount of water in motion.

Air motion ranges from near calm to hurricane winds that flatten windmills. In between is a range from about 7 to about 24 miles an hour, the "breezes" of the Beaufort scale. This is the domain of the wind-power engineer.

There are other restraints, of course. It is theoretically possible to recover only something less than 60 percent of the energy in the wind. (Extracting all the energy would mean bringing all the wind to a dead stop, an interesting possibility.) When we subtract inevitable losses that come about because a propeller or turbine is never perfect (and neither is an electrical generator), we end up with something

The Beaufort Scale

Beaufort Number	Air Description	Wind Velocity (mph)
0	calm	0–1
1	light air	1–3
2	light breeze	4–7
3	gentle breeze	8–12
4	moderate breeze	13–18
5	fresh breeze	19–24
6	strong wind	25–32
7	stiff wind	32–38
8	stormy wind	39–46
9	storm	47–54
10	gale	55–63
11	hurricane-like	64–72
12	hurricane	73–82

like 35 percent of the total wind energy recoverable. This is not to be sneezed at, however, for it is close to the efficiency of the best steam-electric power plants. And since the "fuel" is free, the wind generator is that much more attractive.

There is another physical law that applies to wind machines. Kinetic energy, according to basic laws, is proportional to the cube of wind velocity. If a certain speed gives us 2 kilowatts of power, doubling that speed boosts the power to 16 kilowatts. The stronger the wind, the more the power we get from it—until our equipment blows away.

Interestingly, one square meter of vertical area receives on average about 250 watts of wind energy, just about what an equal amount of horizontal area receives on average of solar energy. The wind is thus an alternative source well worth tapping, as our long history with wind machines proves.

The Rise and Fall of Windmills

A long time ago men took to the water. Gradually the paddlers became sailors, first on crude log crafts, and later aboard more effective boats. Just when or how the idea first came that the wind could ease the job of rowing or paddling we do not know for sure. Perhaps some lucky boatman stumbled onto the scheme when the wind caught his garments and propelled him and his craft along so forcefully that he could not miss the discovery. Whatever the case, the Egyptians were sailing ocean-going ships more than 100 feet long as early as 2500 B.C.

The beauty of the wind as a motive force for ships is apparent in the long and successful history of sail. Until a century ago, sailing craft were the standard of the seas. Indeed, in 1853 the clipper ship *Sovereign of the Seas* once sailed a distance of 485 miles in a single day, averaging more than 20 miles an hour, a speed not matched by many present-day engine-powered ships. Robert Fulton and others doomed the old age of sail except as a sport, and now men use coal or oil to drive ships through the water. Nuclear power too has had a turn at powering ships.

Considering the success of wind power on the high seas, it is remarkable how long it eluded inventors on land. It was not until the seventh century A.D. that the first windmills appeared. As we have said, a windmill operates in roughly similar manner to a water mill, the main difference being that air is far less dense than water, and is of course invisible. Both these factors made it a less obvious candidate for a power plant. The Greek scientist Hero wrote of a "wind organ" operated by a "wheel with oarlike scoops, as in the so-called wind motors," but this is the only reference to windmills at so early a time.

Some historians suggest that the ancient prayer wheel, copied in today's pinwheel toys, was the inspiration for the first windmills. Prayer wheels were in common use by about A.D. 400. The first identifiable windmill actually was not reported until about 640. This was in Islam, during the reign of Caliph Omar I. The idea of the harnessing of the wind did not spread rapidly either, and not until about A.D. 950 do historians turn up the next reference. By then windmills operated in Persia, in a desert area that was hot and dry with strong prevailing winds. The Persians apparently first harnessed the wind to pump water for irrigation. Later they used it on the harvested crops, grinding grain. Slowly windmills appeared in much of Europe. They made it possible for the Dutch to reclaim land from the sea and to operate their mills and factories.

This Greek sail-type windmill is still very popular. Located on the island of Mikonos, the mill grinds grain. *Total Environmental Action*

Long before Columbus discovered America, Europeans had also used windmills as air brakes for heavy loads that had to be lowered, and even suggested their use for driving fighting tanks or lobbing beehives onto besieged cities!

It is tempting to think of windmills as having been developed from water mills, but this doesn't seem to have happened. Instead of converting a water mill over a dried-up stream to wind power, engineers seem to have invented the windmill independently. A great variety of windmills were built. One was the classic Dutch type with horizontal drive shaft and huge blades often covered with cloth. Others used a vertical shaft like the Norse waterwheels, with turbinelike blades—including a design remarkably like the currently popular Savonius rotor. From grain mills and water pumps, windmills progressed to sawmills, as in Holland in 1592, and a variety of other mechanical jobs.

Considering the lineage of New York's first settlers, it is not surprising that the city once featured rows of Dutch windmills lining Manhattan Island. The last one burned not fifty years ago. A pair of Dutch mills were built on a bluff that is now part of Golden Gate

Park, in San Francisco, where they pumped water in pre-earthquake days. (After the energy crunch in 1973, citizens began a drive to make them operational again, an effort requiring several hundred thousands of dollars to duplicate work once done for a tiny fraction of that amount.)

Long before New Yorkers operated their windmills, the first American machine had been built at Windmill Point, not far from Jamestown, Virginia. So the windmill has a heritage of nearly four centuries of useful application in the United States. About the middle of the nineteenth century the old windmill brought to America by the Dutch underwent a radical change. A millwright named Daniel Holladay put together a "self-governing" windmill of his invention in his shop in Connecticut. It inspired a succession of further inventions and led to the production of millions of windmills in an industrial revolution that persisted until the 1930s when the Rural Electrification Administration made it unnecessary to rely on the wind any longer for electric lights and power.

More than half a hundred firms manufactured windmills that dotted the landscape and became part of the sound and sight of the land, creaking as they did their chores, or looming motionless against the sky during periods of calm. An estimated 6.5 million windmills were in use in the half-century between 1880 and 1930; it is hard to imagine cattlemen and homesteaders without windmills to pump water. Among the other major users were the railroads, who used the mills to pump water for the locomotives that civilized the West. This was the extent of the wind's contribution to transportation on land. Although a few ungainly "wind wagons" roamed the prairies, the idea was not practical enough to be pursued. Nevertheless, excluding transportation, windmills in the late nineteenth century supplied about one-fourth of power needs in the United States.

The discovery of electricity and the development of the battery and generator led early to the adaptation of windmills to produce this new form of power. In 1894 explorer Fridtjof Nansen of Norway built himself a wind-driven dynamo in the Arctic and had electric lights before this luxury was available even in large cities! Since wind is in plentiful supply in the polar regions, the practice of harnessing it was continued by Admiral Richard Byrd, the noted explorer. His Jacobs wind generator was so dependable that decades later his son found it still working and brought its propeller back home as a memento.

The Jacobs firm began the engineering of wind-operated electric generators in 1925. Early in the design effort, the firm standardized on the three-bladed propeller instead of the two-bladed one, since it gave smoother operation when the windmill tail vane shifted the propeller's direction in changing winds.

The production model was designed to produce from 400 to 500 kilowatt-hours a month with wind conditions encountered in states in the western half of the country. The generator produced electric power in direct proportion to winds ranging from 7 to 20 miles per hour, and such winds were required for two or three days a week. Such a 2,500-watt, 32-volt plant, equipped with a 50-foot steel tower and a 21,000-watt-hour storage battery, sold for a total of $1,025. The windmill alone could be bought for $495. Installed cost per kilowatt of power was thus about $400 for the complete, installed plant. Maintenance costs averaged less than five dollars a year on a thousand installations that were monitored for ten years.

It was the windmill that led, by a kind of ironic logic, to the airplane. We reversed the process of converting wind to power and made mechanical power cause a wind. Indeed, the first crude attempts at powered flight included balloons mounting sizable, hand-cranked *moulinets,* or "little mills." As the airplane grew as a force in our lives, the windmill waned.

A hundred and twenty-five years ago, there were so many windmills spinning away in the United States that they produced nearly 1.5 billion horsepower-hours of work—equivalent to almost 12 million tons of coal. By 1870, wind power was down about half, as steam engines began to compete for the power market. And when Franklin D. Roosevelt created the blessing of cheap and available electricity through the Rural Electrification Administration, he set in motion a mixed blessing, for it sounded the doom of the windmill for pumping and power generation.

The coming of cheap electricity and cheap small engines ended the reign of windmills. Here was power in a more convenient form and

Old-time American windmill, used for water pumping, grain grinding, and other chores.
Total Environmental Action

cheaper in many cases than just the maintenance of a windmill. Only where the wires did not go did the old wind machines hang on. A ray of hope came with the introduction of radio, and Wincharger and Jacobs quickly shifted their windmill production from pumps to electric generators. The Zenith Corporation boosted sales of its radios by offering windmills at reduced prices when a customer bought a radio! An estimated 50,000 windmills brought radio to remote areas. But in the end most windmills clanked to a halt. Of the 6.5 million, only about 175,000 are thought to remain and most of those serve only as landmarks or museum pieces.

One of the many highlights of the Bicentennial Year 1976 was the erection of a large windmill at Windmill Point, Virginia, the site of the first American windmill. It is possible even now to buy a bag of meal ground by an authentic reproduction of early windmills at Newport, Rhode Island, where Jacqueline Onassis and Doris Duke have created a restoration society, or at Williamsburg, Virginia. Out West in Winnemucca, Nevada, there is another relic of the past, a rare horizontal mill that once pumped water but now is a conversation piece generating electricity for a roadside sign.

The European experience was much the same as in the United States. Although windmills provided a sizable portion of Denmark's needs as late as 1910, and even today there are regions like Crete's Lassithon Valley which provide the sight of thousands of brightly spinning cloth-sailed windmills, other machines, powered by other means, took over almost completely.

Russia has done impressive work on wind generators of fairly large size, and once proposed the installation of thousands of such machines all over the USSR. However, American scientists generally report that Russia seems to have dropped such wind schemes in favor of hydroelectric power and other approaches. In fact, one recounted that the Russians he asked about this just smiled and thought he was "primitive." The USSR has ample hydropower sites and plenty of petroleum.

Despite the general fading out of the windmill as a power source, some experimental work continued on fairly large windmills to produce power. Denmark, England, France, Germany, Russia, and the United States all did such work. In 1931, for example, Russian engineers built a huge windmill near Yalta, on the Black Sea. With a propeller 30 meters in diameter (about 100 feet) the mill produced 100 kilowatts of electricity with a wind speed of about 25 miles an hour.

In the early 1950s an Algerian windmill of similar output was tested, and from 1957 to 1968 the Danes operated a 24-meter windmill at Gedser. Mounted on a tower 30 meters high, this plant produced about 200 kilowatts in a 32-mile wind. Germany built a very sophisticated wind turbine at Stotten in the late 1950s. Using a propeller 34 meters in diameter, it produced about 100 kilowatts of electricity in winds of 18 miles an hour.

For all the marvels of wind machines, and despite their excellent record of power-production without burning fuel or causing pollution, there seemed to be no place for them. The world had shifted from wind to more dependable energy; the kind that came in tank cars or pipelines, ready at the touch of a match or the flip of a switch.

There was a catch to that convenience, however. A catch that caught up with us a few years ago.

The Backyard Wind Mechanics

As is usually the case with alternative energy sources, there has long been grass-roots activity with windmills. Articles in mechanics' magazines, and particularly in *Mother Earth News*, have inspired some to turn away from conventional power and embrace the windmill.

The results of thousands of such love affairs with the wind have been mixed. Many would-be practitioners are totally ignorant of the basic physical facts and achieve only failure. It is one thing to have the power turned off at the local utility, but it is something else again to install and operate a 25-foot diameter wind generator to continue the all-electric way of life. For every ten who flunked wind power, perhaps one aficionado made it.

We need a large propeller to intercept a large cross section of wind. A pinwheel held in the hand spins with great rapidity when held out a car window, and the ventilating turbines atop your roof may seem to be producing great amounts of power in a stiff breeze, but power output depends on the area of the propeller. Here again a physical law prevails: the law of squares. If a 10-foot propeller produces 2 kilowatts in a certain wind velocity, a 20-foot propeller will produce 8 kilowatts, rather than only twice as much. A 100-footer will yield 100 times as much as a 10-foot design. And a small propeller will not produce great quantities of electricity. Another problem is the variability of the wind. At times it blows hard, and at times not at all. In a few places it seems to blow all the time, and quite hard. If we design a wind generator's electric innards to make best use of an occasional 50-knot wind, it will be inefficient at 30 knots or 20 knots. And making it function best at the average wind speed hurts its efficiency when gale winds set in. So the engineer settles for designing for the average wind.

A second problem lies in the inconstancy of the wind. For pumping water, it makes little difference whether the pumping is done at noon

One of the Marcellus Jacobs' excellent wind generators being put back in use. No longer in production, these units are considered prizes when found. *Total Environmental Action*

or at 3 A.M. But we generally want our electricity during waking hours, as harassed utilities well know. So a useful wind generator must have some additional equipment to store electricity for use during periods when the wind is not making any power.

The simplest way to store electricity is in a battery, and many old-timers were familiar with the large arrays needed to power lights, motors, and radio when the Wincharger or Jacobs wind machine was motionless. Add to this the need to mount the wind machine a considerable distance above the ground to take advantage of uninterrupted wind flow, and the expense of a wind generator becomes sizable.

All these problems are but challenges for a growing number of independent types who are buying or building their own wind machines. Several commercial models are available, in a range of power outputs from a few hundred watts to several kilowatts. A wind-generator system supplying the power needed for an average electric home will cost about $15,000, and thus take a long time to pay itself off and start returning money to the owner. The answer of wind buffs is to use less electric power. After all, electric heating and cooling is the most expensive and energy-consuming way to do the job. Limiting electric needs to radio, TV, lights, and other necessary appliances may permit a less-expensive wind generator to do the job.

For example, four 100-watt light bulbs, a stereo, a percolator, and a sewing machine consume a little more than 3 kilowatt-hours a day. A 1,000-watt wind generator will produce enough for such austere electric living if the average wind speed is 14 miles an hour. It is obvious that a 200-watt generator, of which many are sold, will not provide much in the way of electrical comforts.

Publications like *Mother Earth News* print many articles by people who have searched out old wind generators or windmills, surplus utility towers, and the like, and by dint of long and loving labor and study put together a machine that converts the breeze that sways trees into honest-to-goodness electricity that will operate radio and lights.

One popular design with the do-it-yourself fan is the Savonius rotor. The invention of a Finn, the Savonius is different in appearance and in operation from the more familiar propeller-driven generators. Looking much like 55-gallon drums split lengthwise and reassembled off-center (because that's what some are), these interesting machines need no tail vanes to keep them headed into the wind. Nor do they need a governor or feathering devices to shut them off in high winds. The Savonius cares not which way the wind comes from, and survives in gales quite nicely. How old the concept really is can be judged by looking at ancient woodcuts of similar design.

There are hundreds of backyard wind generators, from feeble 200-watt plants to that built by Robert Reines of New Mexico, which gives him five kilowatts from one wind generator, with more in two supplemental wind machines. Reines's plastic-dome house is electrified to the tune of lights, hi-fi, tape recorder, TV, refrigerator, electric fan, pumps, mixers, and even an electrically-operated chemical toilet.

Henry Clews of West Holden, Maine, is another user of moderate-size wind generators. In 1971 he moved to a rural location far from power lines and decided to install his own supply in the form of a wind generator. He settled on a 2-kilowatt unit built by Quirks of Australia. With storage batteries, and a small inverter to convert a portion of the windmill's direct current to alternating current for the television set and stereo, Clews spent a total of $2,800, about $200 less than it would have cost him to have a power line run in the five miles from the highway. Clews's first wind generator gave him electricity for about fifteen cents a kilowatt-hour, about five times that of utility electricity but only half what it would cost from a diesel or gasoline engine.

Into wind power all the way, Clews set up a small company called Solar Wind and became a distributor for the Dunlite generator from Australia and the much larger Elektro made in Switzerland. Clews ranks the Swiss Elektro wind generator the best, particularly the 6-kilowatt model which sells for $3,000 on the East Coast. Of course, a tower and storage batteries add considerably to the cost.

There are only half a dozen major manufacturers of commercial wind generators in the world, including two in the United States. These are Dynatechnology (formerly Wincharger) and Precise Power Corporation. The other manufacturers are Aerowatt, of France, and Lubing Maschine Fabrik in Germany. In two and a half years of operation, Clews's Solar Wind Company sold only 50 wind generators, but answered nearly 30,000 inquiries about them, indicating how willing the spirit but how weak the pocketbook.

Clews presently has $10,000 invested in his two wind generators to produce barely enough electricity for modern living (the family cooks with gas, and heats the house with wood). As he points out in his book *Electric Power from the Wind,* "For many people interested in wind-generated power, the price of commercially-available equipment is simply prohibitively high." Matching conviction with action, he has turned over the rights to the Australian and Swiss wind-generator manufacturers, and markets only wind-plant "kits" for the old, original 200-watt Wincharger units, and a newer Sencenbaugh 750-watt, 12-volt DC kit. These sell for $745 and $2,200 respectively.

Before investing a large sum in a wind generator, one should invest the small sum of $7.50 for a hand-held wind meter. With this simple instrument and some patience, it is possible to assess wind conditions at the site being considered. Unless there are "prevalent" winds of 5 to 13 miles per hour, and occasional stronger "energy" winds of 13 to 23 miles per hour, the site probably is not suitable for producing wind-generated electricity.

Another method of learning the wind-power potential of your location is to write to the National Climatic Center, Federal Building, Asheville, North Carolina, 28801, and ask for the "Wind Tabulation, Percentage Frequency of Wind Direction and Speed" at your nearest weather station. Costing $6.50, these charts tell what percentage of the time winds of various speeds blow, and also their directions.

Chicago, long known as the Windy City, turns out to be far down the list of those blessed by plentiful breezes. Fargo, North Dakota, topped a recent listing, based on ten-year National Weather Service readings taken on an hourly basis. Fargo boasts an average wind speed of 14.4, sufficient to do a good job of generating electric power. Wichita, Kansas, was a not-surprising second, but third went to a city seldom thought of as windy: Boston, Massachusetts, which averages

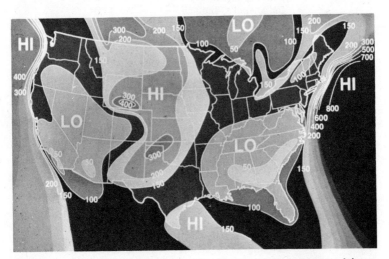

Windmills won't make it everywhere, as this wind map indicates. *Arizona Public Service*

13.3 miles per hour. New York is another surpriser, with a wind of 12.9 and a number four ranking. Honolulu also gets plenty of wind power at 12.1 miles per hour and is the first U.S. city to launch a municipal wind generator. At the bottom of the thirty-two-city list were Anchorage, Alaska, and Los Angeles, California, each with 6.8 miles an hour to make them marginal candidates for wind machines. Washington, D.C., long the butt of wind-power jokes, actually does have a creditable 9.7 miles an hour.

The highest average winds officially recorded in the United States are at Mt. Washington in New Hampshire, where the speed is an attractive 34 miles an hour. This delights the windmill enthusiast, but

Fargo, North Dakota	14.4
Wichita, Kansas	13.7
Boston, Massachusetts	13.3
New York, New York	12.9
Fort Worth, Texas	12.5
Des Moines, Iowa	12.1
Honolulu, Hawaii	12.1
Milwaukee, Wisconsin	12.1
Cleveland, Ohio	11.6
Chicago, Illinois	11.2
Minneapolis, Minnesota	11.2
Indianapolis, Indiana	10.8
Providence, Rhode Island	10.7
Seattle–Tacoma, Washington	10.7
San Francisco, California	10.6
Baltimore, Maryland	10.4
Detroit, Michigan	10.3
Denver, Colorado	10.0
Kansas City, Missouri	9.8
Atlanta, Georgia	9.7
Washington, D.C.	9.7
Philadelphia, Pennsylvania	9.6
Portland, Maine	9.6
New Orleans, Louisiana	9.0
Miami, Florida	8.8
Little Rock, Arkansas	8.7
Salt Lake City, Utah	8.7
Albuquerque, New Mexico	8.6
Tucson, Arizona	8.1
Birmingham, Alabama	7.9
Anchorage, Alaska	6.8
Los Angeles, California	6.8

the dream is highly impractical. Mt. Washington winds often exceed 100 miles an hour, a distinct hazard to any windmill in its path. Herein lies the biggest problem with wind, for it is generally the strongest where people aren't. It is attractive for remote dwellings, but all the movement in recent years has been in the reverse direction, to the constantly bigger cities where one cannot mount in a 25-foot propeller atop a 50-foot tower.

The greatest hope for wind power lies, according to some experts, in a 15,000-mile area centered about fifty miles east of Amarillo, Texas. Certainly the winds blow robustly in the Plains region and in a few other parts of the country. Someday we may be able to tap the wind there and pipe its electrical output to where the people are.

It is obvious to many people, including some who were once ardent wind enthusiasts, that wind power is not for everyone and that we will never see a windmill on every roof. Barring some unexpected breakthrough, the task of furnishing wind-generated electricity on an individual basis is a tough one and will succeed only in certain circumstances. There is a school of thought that says it is unwise to hope even for a modest number of such installations, suggesting instead that if wind power is to be a factor in our lives it must be done on a large-scale, centralized utility basis. As it turns out, there has been a wealth of research and development done in that category as well.

Putnam's Giant Windmill

The largest windmill yet built in the world was that designed by Palmer C. Putnam and financed by the S. Morgan Smith Company. Putnam was an engineer who fell in love with the wind in 1934 and tried to find a suitable wind generator to produce electricity for his home on Cape Cod. Commercial models seemed too small, so he began an investigation of larger machines, turning to Russian, German, and French work. He learned that the Russian Black Sea windmill transmitted power twenty miles to back up a steam plant in Sevastopol. In a single year it was said to have generated 279,000 kilowatt-hours of electricity.

Putnam's interest quickly shifted from domestic windmill applications to the idea of building a giant machine that could feed commercial amounts of electricity into a utility grid. He was able to convince the S. Morgan Smith Company that they should build such a windmill, and he also sold the Central Vermont Public Service Corporation on the prospect of using wind-generated electricity to back up its hydroelectric power to handle peak loads.

Putnam and his group surveyed half a hundred Vermont hills in search of a proper site for the big machine and selected one near Rutland, named Grandpa's Knob. It was about 2,000 feet high, high enough for good wind yet low enough to escape severe freezing conditions. Among those advising in the design of the big machine were Dr. Vannevar Bush, better known for his wartime work on radar and also the atom bomb, and the aerodynamics genius Theodore von Karman. An ardent supporter was Thomas Knight, vice president of General Electric. A sailing enthusiast, Knight liked the idea of harnessing the wind to such good purpose, and he enlisted the aid of both MIT and Caltech for the project.

Indicative of the careful nature of the advance research was wind-tunnel testing using models of the chosen site and rigorous meteorological research. The wind generator itself was a marvel. Even today, the Putnam design is considered excellent. Its large blades were of stainless steel skin over ribs of the same material. The blade pitch could be changed during operation for various wind speeds, and both blades could be feathered, or pointed directly into the wind when it became too strong for safe operation. The Budd Company, famous for its streamlined trains, built the propeller. The 110-foot tower was constructed by the American Bridge Company.

After many weeks of careful testing, the big generator at Grandpa's Knob began producing electricity to feed into the utility lines on October 19, 1941, just a few weeks before Pearl Harbor triggered war with Japan. For three and a half years it would operate, at a speed ranging up to 70 miles an hour. Blades feathered, it withstood gales of 115 miles an hour. It was designed for a maximum output of 1.25 megawatts and at times produced 1.5 megawatts.

Rushed to completion for fear that war priorities would stop the project, the blade spars were not as strong as they should have been.

Palmer Putnam's marvelous 1.5-megawatt windmill built more than 35 years ago. *NASA*

Each blade weighed 8 tons, and stresses during rotation were high. Thus the project was proceeding on a calculated-risk basis. Nevertheless, three and a half years of operation were completed with no incidents. A shutdown in 1943 occurred because of a burned-out main bearing, and it was two years before another could be provided. In early morning darkness on March 26, 1945, after three weeks of operation with the new main bearing, a propeller blade suddenly failed.

Eight tons out of balance, as the broken blade arced 750 feet through the darkness and plunged into the ground, the remains of the windmill vibrated like a giant gone berserk. The foreman, atop the control car, was repeatedly thrown to the floor and only with great effort was he able to reach the controls, feather the remaining blade, and stop the machine. And that was the end of the most ambitious windmill yet built.

In 1939 Putnam and others had estimated that in northern New England, a battery of ten 1,500-kilowatt wind turbines could produce electricity at $.0025 per kilowatt-hour. In 1945 this was revised to about $.006 per kilowatt-hour, and about $195 per kilowatt installed. However, Central Vermont Public Service Corporation could only afford $125 per kilowatt. Although some preliminary studies suggested ways to reduce the cost so that it could match that of fossil-fuel power plants, the S. Morgan Smith Company regretfully decided it could not afford to risk further expenditures on top of the millions already spent. The Grandpa's Knob experiment had demonstrated that a windmill could pump electricity into utility lines. But economic feasibility had not been demonstrated.

Putnam located a wind-power site at Lincoln Ridge in Vermont which he claimed had a capacity for producing about 50,000 kilowatts. He also guessed that worldwide there might be fifty such sites, suitable for power production, and close enough to load centers to justify their construction. In addition to the northeast United States, he mentioned Newfoundland, the Maritime Provinces of Canada, Iceland, northern Ireland, Scotland, Scandinavia, Chile, Tasmania, New Zealand, and possibly the Italian Apennines. The total of all these would come to about 2,500 megawatts, an output exceeded by a number of single hydroelectric power plants and approached by many fossil- and nuclear-fueled plants.

In the decade that followed Palmer Putnam's brilliant accomplishments, the federal government came tantalizingly close to doing something about wind power. The Federal Power Commission was well aware of Putnam's work because the Central Vermont Public Service Corporation included the plant in its 1942 report as an electric utility. The FPC was responsible for looking at new energy sources, and it gave engineer Percy Thomas the task of surveying the wind as an alternative to conventional sources. Thomas was equal to the job and he spent more than a decade on the project, writing his last paper on it in 1954, five years after his retirement from the FPC. In all, Thomas wrote four papers on wind power: "Electric Power from the Wind," March 1945; "The Wind Power Aerogenerator—Twin-Wheel Type," March 1946; "Aerodynamics of the Wind Turbine," January 1949; and "Fitting Wind Power to the Utility Network," February 1954.

Thomas first studied wind data in the United States. Next he designed a wind-power electric plant to produce 7.5 megawatts, some six times as much as Putnam's design, and which he thought could be built for $68 a kilowatt. Finally he studied ways of integrating wind power into the existing utility networks. Convinced that wind could

Percy Thomas's design for a 7.5-megawatt windmill mounted hundreds of feet in the air. *NASA*

be a winner, Thomas strongly recommended to the FPC that a program be undertaken to do something about it.

The record from 1945 onward is vague. A hearing was finally held in 1951 at which the FPC and the Department of the Interior agreed enthusiastically to get on with the design and construction of a wind generator (which had grown now to 7.5 megawatts output) and tie it into an existing hydroelectric power plant. But nothing was ever done beyond the first glowing agreement.

Another voice was heard advocating the large-scale use of wind power to boost our electricity supplies. In 1952 Marcellus Jacobs, whose firm had built and sold some $50 million worth of wind generators over several decades, suggested mounting moderate-size wind generators atop large power lines and letting them feed power into the utility grid as it was produced. He envisioned 1,000 wind generators perched atop transmission lines from Minneapolis to Great Falls, Montana.

In his last paper on wind power, Thomas made no reference to any follow-up of his recommendations, and apparently that was the end of federal interest in the wind until very recently. There are a number of factors that would have put off work on a windmill project of this magnitude. The Korean War commanded the nation's attention. Wind energy was undependable when compared with the still cheap coal, oil, and gas fuels available. And nuclear power was being hailed as the answer to all our energy needs, present and future. Almost two decades of the atomic age would pass before the federal government again turned its attention to wind power.

Federal Effort

It is remarkable that the U.S. government, or any government agency, for that matter, did not become involved in wind-power research until 1972. In that year the National Science Foundation and NASA jointly organized a Solar Energy Panel to assess that resource for its potential contribution to the nation. The panel promptly and surprisingly reported that there might be enough wind energy in various parts of the country to provide 19 percent of electric power needs by the year 2000! A major recommendation of the panel was that the federal government get busy implementing programs to harness the power of the wind.

NASA had begun its wind-generator research in 1972, responding to a request by Puerto Rico that it develop such a machine for possible use there. In March 1973 there was an NSF/NASA workshop on wind energy. Wind buffs, government representatives, utility people, and scientists were invited. Topics considered included past work with windmills, wind availability and siting considerations, design of efficient windmills, energy conversion and storage, and actual design of wind-power systems and the towers they would need. Wind potential was reviewed: 400 billion kilowatt-hours along the Aleutian Chain off Alaska, more than 300 billion kilowatt-hours off New England, 210 billion in the Great Plains, 190 billion off the Texas Gulf coast, and 180 billion off the Atlantic seaboard.

Some thirty years after Palmer Putnam's giant windmill was built in Vermont, then, a joint wind-energy research and development program was initiated by the National Science Foundation and NASA's Lewis Research Center at Sandusky, Ohio. The reason for the project of course was the onset of the energy crisis. Putnam's design had missed economic feasibility by only about 20 percent, and with the price and availability of fuels becoming ever more questionable, researchers decided that the wind's time may well have come.

The NASA-Lewis Research Center at Plum Brook, Ohio (near Sandusky), got the job of designing and building the experimental wind turbine. In preparation for this work, NASA engineers erected a 200-foot meteorological tower and measured wind speed and direction at four heights ranging from 30 to 195 feet. Small, commercially available windmills were also tested, including the French Aerowatt 4.1 kilowatt wind-turbine generator.

While the power output is less than a tenth that of Putnam's machine, the NASA design is almost as large as the earlier windmill. The 100-kilowatt machine mounts two aluminum blades, each 62.5 feet in diameter, on a tower 125 feet high. Each blade weighs one ton. The machine has no tail plane as used on earlier windmills, but copies many European designs which place the blades downwind of the housing, rather than into the wind.

The NASA generator begins to produce electric power when wind

The 100-kilowatt windmill built by NASA at Plum Brook, Ohio. *NASA*

speed reaches 8 miles an hour and attains its rated 100-kilowatt capacity at 18 miles an hour. According to estimates, an average annual wind velocity of 14 miles an hour will produce enough electric power for 30 average-size residences. However, Plum Brook has average winds of only 10 miles an hour, somewhat less than the 14-mile average NASA mentions.

At last, under the auspices of the new Energy Research and Development Administration rather than the NSF, the NASA 100-kilowatt windmill began operation late in October of 1975. Its cost was about $650,000. In a year of operation it is estimated that it will produce about 180,000 kilowatt-hours of energy. At five cents per kilowatt-hour, this represents $9,000 worth of electricity, and it would thus take more than seventy years for the investment to pay off, assuming no costs for operation and maintenance. However, estimates also show that while initial cost was $6,500 per kilowatt, in production this figure would probably drop to about $1,500, giving a payoff in about eighteen years.

The Windmill Forest

Even one windmill 100 feet high and spinning 125-foot blades seems a monster machine. But there are dreamers who suggest building 300,000 far larger wind machines than the NASA prototype. In 1970 Dr. William Heronemus (the scientist involved in sea thermal energy research) began studying the wind along with other alternative sources of large amounts of electric power. Soon he had conceived an "offshore wind-power system" anchored off New England in an area of strong and steady winds, a scheme that he said would produce 160 billion kilowatt-hours of electricity per year. Although cost estimates were slightly higher than conventional electric power, Heronemus pointed to the benefits of no pollution and the saving of scarce fossil fuels for better purposes.

The windmills of Long Island would straddle tall "Texas towers," or be mounted on huge floating barges perhaps 200 feet by 500 feet. Patterned on Putnam's pioneering giant, Heronemus windmills would be built in clusters of six on each tower or barge platform. A total of 640 platforms would substitute for a proposed nuclear plant in Shoreham, New York.

The larger complex of 300,000 wind machines would be built on land, perhaps in the windy Great Plains area. In both cases, they would tie into existing utility lines to provide electricity for consumers. One bonus, according to Heronemus, would be that failure of one or even several windmills would have no catastrophic effect on the system, whereas a failed nuclear or conventional steam-electric plant can cause widespread blackouts and all the chaos that goes with such events.

Decades ago, windmills were a common sight on the Great Plains of this country, pumping water where no other power source existed, and later generating an appreciable amount of electricity. It is here, stretching from Texas to the Dakotas, that Heronemus envisions his great forest of mills, set in clusters of twenty on monster 850-foot tall towers. Perhaps one tower per square mile would be the spacing, but seen from a distance it would seem a forest of whirring blades. Such a colossal array would produce something like 200,000 megawatts of installed capacity, equivalent to 200 nuclear or fossil-fuel power plants. For comparison, U.S. total capacity in 1970 was only about 360,000 megawatts, so the wind forest would be more than half that amount.

Built in sufficient numbers, the windmills would cost about $100 per kilowatt, according to Heronemus. This is only a fraction the cost of any other kind of power plant. However, it remains to be seen how environmentalists—and workers in nearby fields—will take to structures as tall as 100-story buildings and whirling busily as beehives with their twenty giant propellers each. It does not help to recall the failure of the propeller on the original Putnam machine, and its 8-ton blade hurtling through the air.

Heronemus also suggests suspending large numbers of wind generators on cables strung between towers. One proposal for the state of Wisconsin included a series of what Heronemus called "wind barrages," batteries of wind generators mounted in "cage-mast" towers from 100 to 600 feet above the highways. Instead of the huge propellers used in other concepts, these would be only 30 feet in diameter, producing about 20 kilowatts each.

When someone raised the question of the hazard huge windmills might pose to gliders and airplanes, Heronemus unhesitatingly replied that it would be up to pilots to be on their guard; surely the

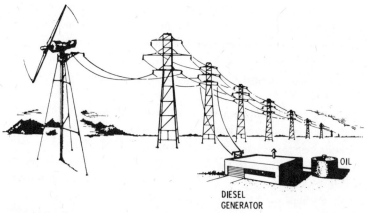

Large windmill used as supplement to conventional electric power plant. *NASA*

need for power is so vital that freedom of the skies is a small price to pay for it.

Environmental Effects of Windmills

Serious study has been given to the possibilities of various kinds of "pollution" by wind generators. Esthetic pollution is the most obvious, and the vision of hundreds or thousands of monster wind machines strikes terror or anger in the hearts of many environmentalists. There are other more tangible problems, including interference with television reception, and possibly even a change in the environment by affecting the normal flow and circulation of the winds. However, one committee suggested that perhaps many farmers would appreciate great batteries of machines that would not only produce electric power but also "soak up the wind" in the process.

Seldom considered except by those deeply involved in such programs are the legal considerations of putting up a huge machine that may steal the wind from a downwind neighbor, affect the weather, environment, or TV reception in the area, or throw a blade and cause damage or injury.

Blowing in the Wind

At a Congressional hearing on wind energy in 1974, Marcellus Jacobs, the pioneer in windmill design and construction, remarked on the sudden hurricane of interest in wind power: "From our observations, which extend all over the country, we have come to the conclusion that there is presently a lot of money and engineering talent being wasted in the sudden drive to get on the 'Wind Energy Bandwagon.'" Wasted or not, the money and engineering keeps flowing in pursuit of successfully harnessing the wind, or at least cashing in on the important sums of federal money that have been pouring out of Washington for some time.

As loud as the cries for building giant wind machines some of the shouts are against them. And even those who advocate wind energy are in many cases dead set against letting scientists and airplane companies do the job. The feeling is that the windmill was invented and fairly well developed long ago and that there is no need for further expensive and lengthy research. The fabulous costs of some experimental wind projects bear out this warning. Pioneer windmill manufacturers have now been supplanted by large aerospace firms including Grumman, Boeing-Vertol (drawing on its helicopter expertise) and Kaman (doing the same thing). General Electric, of course, was involved in the Putnam windmill project at Grandpa's Knob.

The traditional windmill mounts from two to a couple of dozen metal blades. While it might look like engineering wisdom to put as much metal in front of the wind as possible, the real reason for the multiblade approach is to give the high-starting torque required to pump the water most windmills were installed to take care of. For highest efficiency, two- or three-bladed machines are built for generators. However, there are always those who don't get the word on such matters, or if they do they are not convinced.

American Wind Turbine of St. Cloud, Florida, has designed a 15-foot wind wheel, complete with a "bicycle tire" around its multitude of blades. The result is a far cry from the lithe two-blader at Plum Brook, or even the workaday Wincharger or Dunlite units so loved by *Mother Earth News* readers. Weighing a trim 70 pounds, rather than the usual several hundred for a 15-foot diameter windmill, the design offers not only high-starting torque, but a claimed efficiency of 50 percent, which puts it in the ballpark with just about anything else. Also, the ring around the blades eliminated complex gearing mechanisms necessary with conventional windmills. The University of Oklahoma is testing the design exhaustively.

A variation of the wind wheel was put together by Edmund Salter of San Diego, California. This Wind Power Systems RD-7000 turbine consists of three-bladed, rimmed turbines, each about 8 feet in diameter and mounted in a cluster. The rims drive the electric generator centered within the cluster. Another concept, hardly a newcomer, is the Princeton "sailwing," a windmill mounting blades made of a metal spar but with a loose sail of cloth. Such floppy mills, resembling the picturesque ancient Greek machines, are surprisingly efficient, although it is problematical if they will be long-lived enough to be economical.

Eggbeaters and Other Strange Designs

The most sophisticated and glamorous windmill presently on the scene resembles nothing more than a 15-foot eggbeater reaching up into the sky from atop a building at Sandia Laboratories in Albuquerque, New Mexico. At first glance it does not seem to be a windmill at all. This is because it is a vertical turbine, rather than a conventional horizontal-shaft machine. Additionally it does not look as if it could convert the wind that spins it into useful amounts of electricity. Yet its backers see it as a revolution in wind generators.

For all its futuristic look, the vertical-axis turbine is actually a composite of two ideas long ago dreamed up by wind-energy engineers. One dates back more than a thousand years, although it was resurrected and patented in the 1920s by Savonius. The other was designed in 1925 by a Frenchman named G. J. M. Darrieus. Listed in the patent as "having its rotating shaft transverse to the flow of current," the Darrieus turbine, or hoop, was rediscovered independently in the early 1970s by workers at the National Aeronautical Establishment of the National Research Council of Canada.

The Darrieus hoop qualifies as a bona fide invention in the centuries-old history of wind machines. Its inventor gave it the name "jump rope" for the shape of its vanes as they turned, and the device has been scientifically designated a *troposkein,* Greek for "turning rope."

The advantages of the Darrieus hoop include the ability to accept wind from any direction, simpler tower construction, and lower costs of fabrication. Yet it reportedly extracts as much power from the wind passing through it as does a conventional windmill. The design has but one defect: a very high starting torque. To eliminate this difficulty, engineers have added small Savonius rotors at the top and bottom of the hoop. These motors start quickly in the faintest wind, and thus help get the larger Darrieus hoop spinning at its maximum efficiency in less time.

Besides the conventional bladed wind generators, there are a number of more sophisticated approaches intended to produce more power for less money invested in the wind machine. The simplest of these is the tapered-duct approach, in which a large opening funnels

The Darrieus hoop windmill, with Savonius rotor added for easier starting.
Sandia Laboratories

down before reaching the blade. This results in more wind energy striking the blades, at an expense only of the tapered duct.

A more exciting variation is the concept of heating a large area with solar energy to create a rising current of air, then tapping the wind energy of the masses of air that rush in to replace the rising air.

This idea was suggested in 1973 at the NSF/NASA Wind Energy Workshop by a La Jolla, California, scientist. More recently, Grumman engineers have proposed a somewhat similar idea which has been nicknamed "the artificial tornado."

Brainchild of James Yen of the Grumman Aerospace Corporation, the novel wind generator would consist of a vertically-mounted tube, perhaps 20 meters in diameter and 60 meters tall. Air admitted through slots facing the direction of the wind would assume a rotating motion and greatly reduce air pressure at the center of the tube. This reduced pressure would create a vertical wind of very high speed. According to Yen, such a tornadic turbine would be more efficient than the conventional windmill because of its principle of operation. He suggests a power output of one megawatt from turbine blades only 2 meters in diameter, as compared with the 65-meter diameter required for conventional windmill blades to produce that amount of power.

There are other unorthodox concepts for producing power from the wind. One of the most interesting is the "Magnus effect." When a cylinder is rotated, and wind blows across it, a low-pressure area is created on one side of the cylinder. A German engineer built such a four-rotor device in 1926. It was 65 feet in diameter, on a tower 108 feet high, and reportedly developed about 30 kilowatts in a 23-mile-per-hour wind. According to Theodore von Karman, however, it was very inefficient.

Decades ago, crude "wind wagons" lumbered across the Plains in a dry-land copy of sailing ships. Never a real success, such vehicles have long since become museum pieces. Now, however, scientists at Montana State University are rediscovering the old wind-wagon concept, putting it on an oval track, and hoping to produce electric power from the strong and steady Plains winds.

Funded by the NSF, the project is directed toward tracked wind-driven vehicles that would produce from 10 to 20 megawatts of electric power. Rather than the old cloth sails of the wind wagon, the wind train would mount rigid airfoil sections that would be automatically adjusted to derive the most power from the wind as the train proceeded—on wind power, of course!—around the five-mile track.

There are problems in phasing wind-generated electricity into utility grids (Putnam was lucky to have hydroelectric plant storage in the Vermont experiment), so a number of alternative ideas have been dreamed up. Why not produce hydrogen with the electricity, and store the hydrogen for use when needed? Or why not make fertilizer for farm use? Much fossil fuel is used for this purpose, so the idea seems a good one. Dr. Heronemus has a government grant to study ways to heat homes with windmills. The hydrogen idea is not as new as it might sound. Back in the 1890s Danish engineer Poul LaCour used wind-generated electricity to produce hydrogen and oxygen, then burned the two combustible gases to produce a lighting system for a high school where he taught.

New Breed of Sailing Ships

One of the pleasant surprises in the wind revolution is the prospect of a new generation of ships powered by the wind. It would be poetic if the earliest human use of wind should also be the newest. Dr. Homer Stewart, one of those involved in the large Putnam windmill, has calculated the power generated by the wind in driving sailing ships. The last generation of such vessels, built in the late nineteenth century, developed something like 13,000 horsepower per ship. To prove that the sailing ship is not a relic of the past, 200 such vessels assembled in Newport, Rhode Island, at the end of June 1976, and sailed from there to New York to celebrate the Bicentennial on the Fourth of July.

Meantime, in Germany, an up-to-date version of the sailing ship is said to be under construction. Designed by a civil engineer and aerodynamics expert named Wilhelm Prolss, the "DynaShip" will resemble conventional vessels in its hull. However, its four square-rigged 200-foot sails will have no rigging and will be controlled automatically by remote control from the bridge. Instead of wood the yards will be thin stainless steel, and the sails Dacron.

The proposed DynaShip would be a 17,000-ton freighter, 400 feet long. Computers would calculate sail settings, and the ship would sail lanes of best winds as reported by satellites and other sophisticated weather reporting. Sailing at 20 knots under favorable breezes, the

The DynaShip design is hoped to return wind power to the sea lanes. *DynaShip Corp.*

new clipper would get about 95 percent of its energy from the wind and would use its auxiliary engine only when necessary.

Fair Wind for the Future

It has been estimated that about 2 percent of solar energy is converted to wind energy in the atmosphere. Further calculations based on this estimate are that the wind energy over the United States is about fourteen times the present rate of consumption of all kinds of energy. It would be ridiculous to think of harnessing all the wind, but a study by the NSF and NASA recently estimated that 1.5 billion megawatt-hours of wind power might be produced annually by the end of the century. This would equal total U.S. electric power use in 1970, and thus go a long way toward filling the demand for power.

The Federal Energy Administration also has looked long and hard at wind power and has evaluated it as sizable, nondepletable, free of foreign problems, with few environmental effects, not requiring water, and with a small land-use requirement of between ¼ and ½ acre per megawatt of installed capacity. An estimated power of 45 megawatts per square mile might be available if grids of wind generators were constructed in the windy regions of the country.

The FEA's Project Independence reports suggested two scenarios for wind potential. On a "business-as-usual" basis, utilization of windmills for electric power ranged from 400,000 megawatt-hours a year in 1990 to 300 million megawatt-hours yearly by the year 2000. If an accelerated program were carried out, the numbers would increase to 2 million megawatt-hours in 1980 and 15 billion megawatt-hours yearly in 2000, ten times the NSF/NASA prediction.

A key to successful harnessing of wind power is the cooperation of the electric utility companies, of course. Aware of this fact, NSF/NASA sent out letters to 2,700 utilities in the United States in 1974. Of 382 replies received, 68 expressed strong interest, 251 some interest, and the remaining 63 no interest at all. Hardly a mandate for developing wind power, but at least one utility in ten is testing the wind with some interest.

7. Biofuels: The Renewable Supply

We depend on energy—in the form of food—to sustain the most vital of our activities. The sun provides these foods, or biofuels, that keep us alive, and fossil fuels are in a sense stored biofuels, for they were living organisms eons ago. Before we stumbled onto fossil fuels and made them the basis for our energy economy, we depended on biofuels not just for food but for heat, light, and power as well. Wood, grown by the photosynthetic process which converts solar energy, water, and carbon dioxide into carbohydrates, has been burned for fuel ever since man first claimed the gift of fire. Wood still serves many people, including some in our own country.

Other biofuels are in fairly common use. Peat is one example, animal wastes another. Just as advanced societies burn up oil, gas, and coal resources which could better serve other purposes, so the less-advanced countries burn up animal dung that would be far better used as fertilizer. As we near the bottom of the fossil fuel bin, it is natural to seek a continually renewable supply of biofuels. Since we can't wait millions of years for new coal or oil, why not use biofuels on a current basis? Such an approach draws on "income" rather than "capital" energy.

Peat is a biofuel just starting out on the long journey to become coal. Peat is not used in the United States, but Ireland, Russia, and Germany use it to some extent. In total, about 25 million tons are burned as fuel and to make coke. Perhaps another 8 million tons worldwide are used for other purposes. This is an insignificant amount of fuel, but world reserves of peat are estimated at 136 billion tons. It is possible that with proper management peat bogs could be made to yield more than a billion tons a year for an indefinite period.

Carbon is the common denominator of biofuels. Coal, oil, and gas are largely carbon, so are wood and other vegetation that can be burned to produce heat energy. The renewable biofuels include wood

Coal is stored-up biofuel. *Bureau of Mines*

and other burnable biomass; methane and other gases; organic wastes and liquid fuels. Biofuels differ from the other forms of solar energy in that the storage problem has been solved. This is what makes them worth the increasing attention they are receiving.

Food supplies body heat and the energy for movement and doing work of various kinds. When we added domesticated animals as sources of food and also energy, solar energy fueled these too. The sun furnished the fuel for cooking when we added this refinement to the eating process. Wood fires later did many other chores, from industrial processes to the production of mechanical energy in steam engines. We were well on our way to an extensive renewable solar-energy economy when we discovered the treasure trove of peat, coal, and finally gas and oil.

Our present dependence on fossil fuels is a very recent development in civilization. While we now are so tied to coal, oil, and natural gas that we cannot do without them, they came on the scene only yesterday in the scale of things. It was about 1800 when Americans began to substitute coal for wood. Oil was discovered in 1859, and took decades to become a great factor. Natural gas is truly a Johnny-come-lately, and was put to wide use only in 1930! Once they became entrenched, however, these convenient fuels quickly displaced almost everything else. Wood, which once was our mainstay, now provides no more than 1 percent of our energy needs. Yet ironically we are now giving wood a serious second look.

Photosynthesis

Basic to the magic performed by the sun is the process of photosynthesis, the putting together of certain organic substances with radiant energy. Human beings and animals are not energized directly by sunlight but indirectly by plants and by animals that have refined the content of plants *they* have eaten. Plants are not carnivorous, neither are they vegetarian—instead, their energy comes directly from the sun. In photosynthesis, sunlight "fixes" carbon out of the atmosphere. It needs the help of the miracle compound chlorophyll to do this. This stuff that makes grass green acts as a catalyst, or mover and shaker, to get things going in the photosynthetic process.

A manmade solar collector relies on the infrared, or heat portion of the solar spectrum. In photochemical processes, however, infrared is of no use in storing energy. A plant makes use only of visible and ultraviolet light. The mechanism of photosynthesis is very complicated and is only beginning to be understood. What seems to happen is that photons of light striking molecules of carbon dioxide and water activate those molecules so that their constituents are re-shuffled and synthesized into carbohydrates. Nature does miraculous things with incoming solar energy. Sugar, which is available immediately for energy, is produced in fruits and plants. For short-term storage, starch is produced, which can be converted to fuels with little difficulty. Over a longer period, cellulose is produced, providing long-term storage of energy. The stuff produced by the photosynthetic process is sometimes referred to by the inelegant name of biomass, and this term is generally extended to animals and human beings.

Making fuel from sunshine depends on a variety of factors, one of them being the amount of solar energy available to make carbohydrates from carbon and water. For example, an Iranian might produce 8 tons a year per acre in his sunny farming environment, while a Hollander would be limited to a little less than half as much, because on average he gets less than half as much solar energy. His high northern latitude makes the sun rays strike the ground at an angle rather than from overhead. Holland also has fewer hours of sunshine.

An acre of farmland or forest producing 2 or 3 tons of dry foodstuff or other plants per year represents an energy conversion efficiency of only about two-tenths of one percent. Solar cells make electricity from sunlight at an efficiency rate of up to 20 percent, and green plants look hopeless as fuel sources by comparison. Nevertheless, because of the vast potential in bioconversion, even at low efficiency, we are taking biofuels seriously.

There are also ways of fooling Mother Nature as she carries out the photosynthetic process, or rather, of helping her do it more efficiently. Since the entire surface of a field is not filled with crops, much of the sunlight goes right on by without a chance to be converted into biomass. The sun also strikes the field at an angle for much of the time, and half of its light is not the right kind to be used by plants. Later in this chapter we shall look at some of the methods of improving the situation and producing considerably more fuel per acre of sunlight.

The "Energy Ranch"

Although a much smaller population once relied on wood for fuel, it seems as impractical to return to a wood economy as it would be to shift from jet airliners to the Wright brothers' aircraft. Nevertheless, some energy experts are suggesting the "energy ranch" concept: instead of food or fibers, the farmer grows energy for sale to the consumer. The energy ranchers talk seriously of "Btu bushes" of various kinds, from such obvious candidates as pine and certain hardwoods to fuel crops including sugarcane, bamboo, Bermuda grass, water hyacinths, and kenaf.

Most solar energy falls on water rather than land, since water covers more than 70 percent of Earth's surface. However, the sun falling on land fixes or reduces about 16 billion tons of carbon a

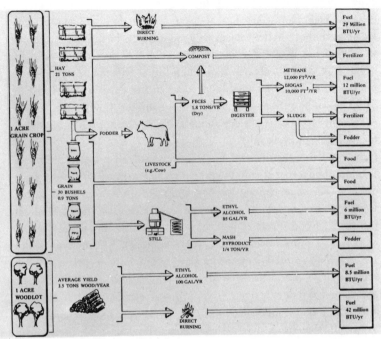

The various outputs possible from an "energy ranch." *Energy Primer*

year. This compares with about 3.5 billion tons of coal produced worldwide in 1970. To the bioconversion enthusiast, the temptation is great: here is the potential for more than four times our coal resources. The catch is that if we burn *all* the carbon produced by trees and other vegetation each year we would have no food, fiber, grain, or wood for other important purposes. The energy ranch must be a compromise designed to provide fuel without threatening that even more important commodity, food.

Wood, the Solid Fuel

Wood has much to commend it as a fuel. It is storable, portable, safe to handle, burns relatively clean without hazardous pollutants, and is relatively compact as well, yielding many Btus per cubic foot. A cord of wood is a pile 4 feet high, 4 feet wide, and 8 feet long. This is 128 cubic feet, but because of the round logs only about 80 cubic feet are actually fuel. Hardwood has about twice the heat content of soft wood, and about half as much as good coal. In fact, a cord of good wood matches a ton of coal, or 200 gallons of fuel oil. A cord of wood may contain almost 25 million Btus. Other fuels provide a million Btu for about one dollar. At $40 per cord, wood comes to $1.60 per million Btus. It is tempting to consider wood as the "moral equivalent" to fossil and nuclear fuels, but there are problems beyond the high cost of firewood.

So great was the demand for wood in England prior to the introduction of coal that wood cutters had all but stripped the countryside of trees. The same denudation had happened on a large scale on the Continent, and people turned to coal largely because they had about run out of wood. Parts of our country today little resemble the land of towering forests the colonists came upon. Then there were about 12 billion acres of timberland; today only 8 billion remain. Try to imagine how much forest would remain had we not mined coal, pumped oil and dammed our rivers for hydroelectric power.

It is true that a lucky few of us are situated near ample supplies of firewood, sometimes free. These fortunates, most of them rural dwellers, escape the energy crunch threatening most of us. Much as we might desire it, it is unlikely that many of us will fill our energy needs with wood. We are presently using about 3 billion acres of forest for lumber, but little of it except scrap for fuel. Were we to try to fuel our economy on wood we would fail dismally while we consumed trees vitally necessary for a host of other uses including homes, furniture, paper, and other products.

Wood is not the only biomass that will burn, of course. Everything from algae to coffee to wheat can burn, and has been burned at sometime or other. During World War I, millions of tons of coffee were burned as fuel in South America when there was no way to ship it to market. Sugar is an excellent fuel, for human beings and for machines. Thus sugar beets have been suggested seriously as a useful energy source—with the idea of converting them into alcohol, an excellent liquid fuel. Grains also produce alcohol. The red tape presently surrounding the production and taxation of alcohol is a problem that boggles the mind and may help discourage such biofuels.

Sugarcane turns out to be one of the most efficient converters of sunshine into biomass. An acre of suitable ground will produce about 20 tons of dry material a year, compared with only about $3\frac{1}{2}$ tons of pine trees. One estimate worked out to provide a graphic picture of annual worldwide biomass production is a pile of sugarcane filling a square field 43 miles on a side and heaped two miles high. This represents more cane syrup than we could make good use of, but a vast potential in energy.

Gasoline from Trees

Wood, which is cellulose, can also be converted into wood alcohol or methanol for use as a fuel or in industry. Sweden resorted to this fuel for its vehicles in wartime. Methanol, long used in cars and trucks, has the advantage of burning cleaner than gasoline but has only about half the energy content. Methanol is often used in clean-air car races.

Millionaire industrialist Robert A. G. Monks, serving as director of Maine's Office of Energy Resources, has pushed hard for converting Maine timber into methanol as a substitute fuel. He points out that Germany still has a substantial methanol technology and that MIT is among those who have proved the feasibility of mixing methanol with gasoline. Maine's interest has kindled a similar spark in neighboring New Hampshire, also blessed with much timbered land, most of it not used.

J. B. Hawley, Jr. of Minneapolis gave $100,000 to MIT to start its Methanol Center for research in this area. Hawley made his fortune in gas and oil but was interested in the development of follow-on sources of fuel as fossils are used up. Dr. Thomas Reed of MIT, director of the Methanol Center, himself worked earlier for Shell Oil Company.

Energy ranch advocates have suggested dense plantings of 10,000 or more trees to the acre. These would be harvested every three to five years, at about 10 to 15 feet in height and less than an inch in diameter. It might be possible to produce from 10 to 15 tons of fuel per acre per year in this manner, and a square mile could supply fuel for an electric power plant producing from 2 to 3 megawatts of electricity.

Bright-eyed planners talk eagerly of individual "plantations" producing enough wood or what-have-you to fuel a 1,000-megawatt electric power plant. Some schemes of this type would need about 650 square miles to do the job, a plot more than 25 miles on a side. A common argument is that present-day pulp mills require similar-sized tracts of forest land to keep them operating, and they are economically justifiable. Yet some energy-ranch advocates admit that an area one-third the size of U.S. farmland would take care of all our needs.

Among the benefits claimed for the energy ranch are a forever renewable supply of fuel, a nationally grown resource that would benefit the balance of payments in international trade, a sulfur-free, nonpolluting fuel (ash is an excellent fertilizer), and no problems of spills or leaks to harm the environment as in the case of oil. The energy ranch would not affect the worldwide or regional carbon dioxide or thermal balance, and it could be planted "next door" to many power plants, thus saving freight costs.

Although the energy ranch has obvious advantages, hardheaded realists feel that even one percent of our energy needs from this method within less than a quarter of a century is wishful thinking. A key consideration must be the economics of growing various kinds of fuels. Grains, for example, have a cash value of only twenty cents per million Btus. So it is difficult to justify using land for wood growing when food brings a far higher price.

Another factor seldom considered by energy ranch proponents is the amount of water needed to grow biomass. Nature is very prodigal with solar energy in producing plant tissue, and also uses great quantities of water in the process. One pound of dry plant tissue requires about 600 pounds of water. Thus 1,000 tons of dry biofuel, a relatively small amount if we are considering operating large steam electric power plants, would require nearly 150 million gallons of water, a commodity that is also in short supply.

What it comes down to is the fact that there probably is not enough suitable farmland on Earth to provide sufficient wood or other biomass to operate the world's machines in the way we have become accustomed to having them operate.

Besides being a delicacy, maple sugar is also a biofuel. So are the trees that produce it. *U.S. Department of Agriculture*

Fuel from the Seas

Converting dry land acreage to the production of fuels would work a severe hardship on production of food and other necessities, so some energy planners are turning to water as a likely breeding ground. A pound of coal of the bituminous variety yields about 13,000 Btus. A pound of *Chlorella pyredenoisa*, a green alga, also yields about 13,000 Btus per pound. And algae offer the potential of much greater conversion efficiency from solar energy to biofuel—some researchers claim as high as 20 percent efficiency, in fact. In the laboratory algae have been cultivated at efficiencies as high as 10 percent, so the higher figure may sometime be attainable.

The research firm of A. D. Little, Inc. of Boston, Massachusetts, has done a great deal of research and development in algae cultivation. So have Japanese researchers. Using solar-heated water, special nutrients, and plastics to enclose the entire growing environment,

Experimental algae production facility atop laboratory building. *Arthur D. Little, Inc.*

A. D. Little workers literally soaked up all the solar energy available per square foot of area and went nature several better in converting sunshine to calories.

Some have suggested that this approach could yield a fuel for power plants. Certainly algae offer higher conversion ratios than wood and other biofuels, but there are great economic problems. A. D. Little estimated some years ago that it could produce *Chlorella* for about twenty-seven cents per pound in large specially-built plants. If people could be convinced to eat the marine foodstuff (and that is another problem), it might make it as a food. But a ton of it would cost $540 dollars, compared with the then cost of coal at $8 a ton. Algae for stoking furnaces and power plants sounds like using gold leaf for house paint!

In addition to algae, plants like the water hyacinth have been suggested as potential sources of fuel. The prolific hyacinth has made a pest of itself by choking waterways, and any attempt to put such a problem to good use should be admired. However, the matter of geography shoots down attempts to shift the energy ranch from scarce real estate to a watery environment. The freshwater area of the United States, even including the seemingly huge Great Lakes, amounts to only about 2 percent of the total. Even with the promise of algae's high efficiency, or the water hyacinth's prodigious growth, these advantages probably cannot overcome the mathematics of the situation. If we can't make the energy ranch go on all the dry land, it will hardly succeed on $\frac{1}{50}$ as much water. Fortunately, we have a much larger area of water to turn to in the world's seas and oceans.

For every acre of dry land there are about $2\frac{1}{2}$ acres of ocean. Water provides only about one-third of the total biomass of Earth, because many stretches of sea are "deserts" and produce little living growth. Nevertheless, turning to the open seas, the bioconversion engineers think there may be possibilities. Remembering the promise of laboratory and pilot pond results with algae culture, some researchers suggest exploiting the algae growth of the oceans. Phytoplankton abounds in the sea sometimes profusely, as in the well-known red tide that occasionally occurs. In the oceans there is space enough and time, and some future engineers may take an appreciable portion of our fuels from the sea, in this biological manner. However, there are problems, including those hinted at by the rampant disease that causes the red tide. Also to be considered are competitors and predators to algae selected for marine farming.

Instead of microscopic algae, marine life on a much larger scale was first selected for ocean experiment. The giant seaweed known as kelp was the choice for the nautical Btu bush. The Marine Farm Project, a joint research project of the NSF, the Naval Undersea Center at San Diego, California, and the Naval Weapons Center at China Lake, California, was initiated in 1973. A modest seven-acre venture, it involved placement of a mesh of lines anchored beneath the surface 60 feet off the coast of Southern California. Seaweed was grown in this manner to ascertain growth rates, yields per acre, length of life of a crop, and costs.

Cost of the pilot operation was about $3 million, and researchers suggested a follow-on phase costing $48 million over a four-year period. This would produce a 1,000-acre marine farm, either in the Pacific or the Atlantic. Success with this extension of the work would lead to a full-scale 100,000-acre marine farm costing an estimated $2 billion and producing appreciable amounts of products and fuel.

Kelp is presently harvested in Japan and the western United States for a variety of uses. In some cases equipment resembling underwater

The water hyacinth is highly thought of as a source for bio-mass to convert into methane for fuel. *U.S. Department of Agriculture*

The net contains plankton, the minute biomass produced in the ocean or lakes. Such sources are attractive to energy prospectors because of the vast size of the ocean. *Bureau of Commercial Fisheries*

lawnmowers cut off the kelp when it is ready for harvesting. The giant seaweeds are convertible into pharmaceuticals, foods, food additives, plastics, synthetic fibers, waxes, and lubricants. When harvested for fuel, the kelp would be dried and chemically processed, perhaps aboard ocean-based plants which would then transfer the processed fuel to tankers for delivery to ports. Despite its barren areas, the ocean potential is obviously great. So interest in marine energy farms is not surprising.

United Aircraft Research Laboratories is among those looking at "ocean farming" for fuel sources. The candidate crop United Aircraft selected is also giant kelp, and the "farmland" would be an area of 250,000 square miles of sea off the west coast of the United States. Starting at the northern end of this watery plantation, kelp seeding would be done in an 80-mile strip running east and west. The slow-moving southern current would then carry the growing kelp to the "harvesting grounds" off the Mexican border in about eighteen months. The estimated yield, with no fertilization of the kelp, would be about one ton of dry organic matter per acre per year. Although this is not a high yield, there would be so many acres that total production would be sufficient to take care of about 2 percent of current United States energy consumption. The estimated cost of kelp fuel per million Btus would be $4.85, several times that of coal.

In addition to its Pacific Ocean tract, United Aircraft's team is considering an even brighter proposal in the Atlantic. Moving to the Sargasso Sea, of Bermuda Triangle notoriety, the marine farmers would seed another large area with kelp and then add fertilizer in "timed-release" form. Because of fertilization, and the warmer water available, the biomass yield from the huge ocean farm would approach 50 tons per acre. Up to 10 percent of U.S. energy needs might be met by such a marine farm, researchers estimate.

We already depend on the sea for much of our food, and some countries rely strongly on fish, seaweed, and other marine life. What effect large-scale kelp or algae farming for energy production would have is not known and must be carefully considered. Dry-land crop production is difficult enough in the matter of pest control, and the problem may well be far more difficult in a watery, three-dimensional environment.

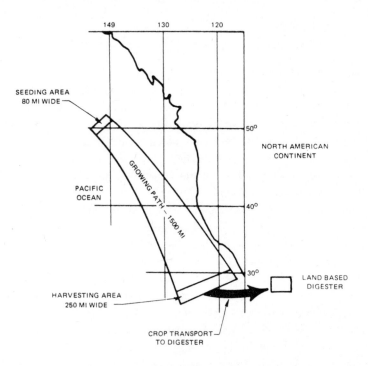

Drawing shows the scope of proposed seaweed harvesting operation off the West Coast. *U.S. Government Printing Office*

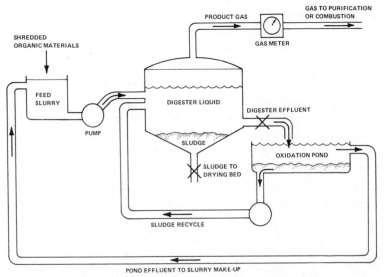

Continuous conversion of organic waste to methane gas by the process of anaerobic fermentation. *Federal Energy Administration*

Methane Gas

The "hydrogen economy" is hailed by many energy experts as the coming of the millennium. This fuel gas produces practically no pollution or residue of any kind, has a very high fuel content, and can be compressed, liquefied, and otherwise manipulated for easy handling and storage. We are some distance from this happy day, but there are gas fuels that have for centuries been a blessing and a problem to humankind. Best known and most used is "natural gas"; this handy form of energy is in very short supply and for that reason many are earnestly and eagerly seeking other sources for it.

When earthquakes struck Nigaata, Japan, in 1952 a large part of the ensuing damage was caused by fires. This was because many households had their own private gas wells, tapping methane produced in marshy ground by the decomposition of plant life. Carried through bamboo pipes, this highly combustible gas played havoc when quakes tore apart the fragile gas-supply system.

The decay, or bacterial digestion, of any organic material produces a number of byproducts, some of them quite useful. There are two basic kinds of decay: aerobic (taking place in air) and anaerobic (taking place in the absence of air). When wastes decay in the open, the result is the production of ammonia, carbon dioxide, and finally the solid called humus. Anaerobic decay yields somewhat different byproducts. In natural anaerobic decay of plant matter under water, peat is produced, along with biogases which escape. In the intestines of animals, a biogas is produced, along with solid waste. Manmade "digesters" of metal tanks and associated equipment also produce biogas and a solid waste called sludge.

Methane itself, chemical symbol CH_4, is an odorless, colorless, highly inflammable gas that is an excellent fuel. It is in fact the main constituent of natural gas. A pound of dung produces about one cubic foot of methane, and 225 cubic feet of this is equivalent to about one gallon of gasoline. One cow in a year produces enough manure to be converted into the equivalent of about 50 gallons of gasoline. Interestingly, although a pound of manure produces only one cubic foot of gas, a pound of dry leaves will make about 7 cubic feet of gas. However, the leaves leave no sludge for fertilizer.

Nature produces methane slowly from waste, but experimenters speed up the process by enclosing the process and adding heat. This enclosure is called a digester because it aids the bacteria in their eating chore. The simplest production method is "batch feeding," in which a load of manure or other biomass is put into the digester and left until all of it has been converted to gas. In "continuous feeding" (usually with liquid waste) the material is fed to the digester over a period of time. A "two-stage digester" is often used in this process, with the first stage producing about 80 percent of the gas and the second stage gleaning the remaining 20 percent. One very worthwhile addition to the methane digester is a solar heater rather than the use of fossil fuels to heat the system.

In 1905 Bombay, India, had a large installation that produced gas and also fertilizer from animal wastes. Germany and England, and to a lesser extent other European countries, produced appreciable amounts of biogas to fill wartime shortages. Algeria, South Africa, Korea, France, and Hungary are among some of the countries using thousands of small digesters. India today is still very big on methane, or "gobar gas," as it is called at the Gobar Gas Research Station at Ajitmal, in northern India. *Gobar* sounds like a good American trade name but actually is Hindi for "cow dung."

There are pressing reasons for India's estimated 2,500 gobar gas plants. The rapidly growing population has all but eliminated forests, and most rural families burn cow dung for cooking and also for heating—hardly an appetizing thought in either application. Actually, this fuel is not merely a nuisance but a serious health hazard. It brings flies, and the acrid smoke can also cause eye disease on a widespread basis. The burning of dung also eliminates its use as a fertilizer, which further aggravates the food problem.

A great variety of Indian methane digesters produce gas in quantities from a modest 100 cubic feet to as much as 9,000 feet per day. In Delhi, methane drives four 400-horsepower electric generators, and the sludge is given to farmers for fertilizer. Small digester designs have been standardized, and construction details have been made available not only to Indian people but to interested experimenters in other parts of the world, including the United States, where there is a great demand from hobbyists. Gobar gas, wherever it is produced, consists of methane ranging from more than 50 percent to 70 percent, carbon dioxide from 27 percent to 45 percent, 1 percent to 10 percent of hydrogen, a trace to 3 percent nitrogen, and traces of carbon monoxide, oxygen, and hydrogen sulfide (the compound that smells like rotten eggs).

In the United States we think of ourselves as a mechanized civilization but we still have some 400 million cows, horses, pigs, chickens, and other animals producing more than 2 billion tons of manure a year. As *Mother Earth News* points out, this is enough to cover a 500-square-mile area four feet deep in manure! Or to produce a huge cloud of methane gas.

Hobbyists have used methane for cooking, driving old cars, running power plants, and the like. *Mother Earth News* has calculated the animal helpers needed to drive your car just fifteen miles a day on methane: 10 ponies, 550 chickens, or 100 human beings. Unfortunately, the gas necessary would fill twenty 50-gallon drums and make quite a load to carry on the car. Compressing it to 1,000 pounds to the inch for more compact storage would use up about 20 percent of the energy one started with and thus require another 2 ponies, 110 chickens, or 20 people. Methane digesters are not yet ready to take the place of the corner filling station, being better suited for cooking, heating, or running stationary engines.

Energy from Waste

Man's history, to hear some tell it, has been one of constantly converting energy to waste. Now the emphasis is increasingly in the reverse direction; turning waste back to energy as the more conventional sources dwindle. Actually, the historical situation has not been as bad as detractors claim. The useful conversion of waste dates far back in time, and many new schemes for salvaging garbage are simply rediscoveries of old ideas. We have generally taken the easy route, however, and the conversion of wastes in most cases has not been that route.

The United States is a great waste maker: each year we produce more than 2 billion tons of organic waste. Just how much energy this represents is a matter of heated debate. Some proponents of waste conversion claimed a few years ago that 2 billion tons of waste could be converted into a tidy windfall of almost 2.5 billion barrels of oil every year. Here was a wonderful thing to consider, for our total consumption was only about 5 billion barrels! The waste converters were guilty of some zealous exaggeration in their eagerness to get a hearing.

More recent and more careful estimates point out that nearly half of the total 2 billion tons of waste is water, which cuts the energy down to about half the original estimate. That would still be one-fourth of our annual oil consumption, but the realists were not through yet. More than 80 percent of the waste was so far from a point of need that it could not be economically collected and transported. This left something less than 150 million tons of dry organic wastes for converting into useful oil fuel. Thus instead of the whopping 2.5 billion barrels the result would be about 170 million barrels. Even taking the low figure, however, the goal seems worth considerable effort, for it amounts to about 12 percent of our imports. About the best we could ever get from *all* waste conversion seems to be about 12 percent of our total energy needs. This of course is not peanuts. Nuclear energy presently provides only one percent of our needs, and projections for the year 2000 show it still only accounting for about 7 percent. So a source, or sources, that can give us 12 percent is certainly worth pursuing.

The 12 percent estimate of energy from waste comes from three major sources: 6 percent from farm wastes, 4 percent from forest scrap, and the remaining 2 percent from domestic waste. Since our farms and forests are widely dispersed, a large part of the useful conversion problem will be transportation. A resource worth $X per ton is not profitable if it costs $2X to get it to a place where it can be used. For the most part, then, use of such diffuse waste energy resources may have to take place at or near the disposal site. An exception might be the conversion of cattle feed lot wastes to methane gas for domestic or industrial use. These facilities concentrate the production of waste and there is much current interest in developing this resource.

The 2 percent potential in domestic waste seems likeliest to be effectively converted to economically affordable fuel supplements.

There are two general sources of such waste. One is sewage, the other is trash and garbage. Because the recovery of sewage gas has a longer history, we will consider it first.

In Exeter, England, in 1895, civil engineer Donald Cameron built the first methane digester used in conjunction with a large municipal septic tank. The gas was used to light the area in the vicinity of the plant. Improvements in Cameron's design were soon made in England and in Germany. Production of gas from sewage thus has a long history and has been brought to a rather refined state. Failure to make more use of this energy source stems from lack of interest, or need, rather than from a lack of knowledge.

Not many years ago, the "flaring" of the natural gas byproduct of petroleum was a common sight in the oil fields. This wasteful practice has largely been eliminated in the United States, although it continues in many foreign oil-producing nations. Even so, with today's tight energy conditions, some producers of methane gas are flaring that valuable fuel. An example is the city of Oakland, which for years has burned the methane gas produced at its sewage plant just to get rid of it. This amounted in recent years to about 500,000 cubic feet a day. The Pacific Gas & Electric Company is now negotiating to buy the waste gas, process it, and use it for fuel. Chicago and Los Angeles produce methane from their sewage plants and use the gas to provide power. It has been estimated that if all sewage in the country was so processed, about one percent of natural gas consumption could be saved.

In addition to bioconversion, there are two general methods of converting dry organic wastes into fuel: hydrogenation and pyrolysis. In hydrogenation, the waste, plus a catalyst like sodium carbonate, is placed in a pressure vessel containing steam and carbon monoxide and heated to a temperature between 240 and 380 degrees C. for up to an hour. Hydrogenation is remarkably efficient, and up to 99 percent of the carbon content of the waste is converted to oil. One ton of waste makes about two barrels of oil. However, it takes some fuel to heat the mixture in the first place, so the net output per ton of waste is actually about 1.25 barrels.

Although hydrogenation at the moment is an expensive process, it is potentially attractive since it does two important jobs at once:

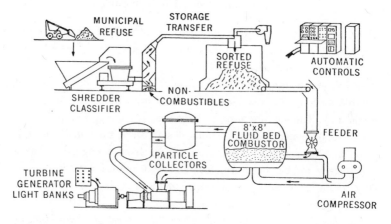

Diagram of conversion of municipal wastes to electric power. *Federal Energy Administration*

disposing of unwanted garbage (which might cost up to ten dollars a ton to get rid of) and producing a product worth about six dollars a ton. It should be noted that there is a pollution problem from the hydrogenation process, since sulfur is given off.

In pyrolysis, at a temperature of about 500 degrees C., the thoroughly dried and shredded waste is converted to gas, oil, or "char." These three different fuels pose problems in handling, and for this reason pyrolysis is not as economically attractive as hydrogenation. On the economic side, estimates are that at a cost of five dollars a ton, waste will produce six dollars worth of fuel, not allowing for credit for waste disposal in the process. Among those working with pyrolysis equipment are the Garrett Corporation, Union Carbide, and Monsanto.

As recently as 1972, we were so confident of our conventional fuel sources that there was not one municipal plant converting garbage to energy in the United States, although this has been done in Europe for years. As this is written, about thirty municipal waste conversion facilities are operating, and another one hundred or so are being built or planned. Although these will by no means convert all the available garbage, they will process perhaps 10 percent of it, a very good start.

One pioneering garbage-to-energy project is a $35-million plant at

Refuse pit stores 6,700 tons of garbage, enough for more than five days' supply for the power plant at Saugus, Massachusetts. *Federal Energy Administration*

Saugus, Massachusetts. Collecting refuse from a dozen communities, the plant will burn garbage to produce steam for use by the General Electric plant at Lynn. This plant amounts to a clean garbage disposal facility for half a million people, and the equivalent of 14 million gallons of fuel oil annually.

Baltimore, Maryland, has a conversion plant that turns 1,000 tons of garbage a day into steam to heat and cool downtown high-rise buildings. As a bonus, the plant extracts 70 tons of metal a day for recycling. St. Louis, Missouri, pioneered in garbage conversion, getting started in 1972 just before the energy crisis broke. It mixes garbage with pulverized coal to fuel a plant producing electricity. Estimated savings are about one million tons of coal a year.

In addition to steam or electricity, some plants yield gas or liquid fuels. In a different approach, the Brooklyn Union Gas Company of New York City is installing a network of collection pipes under a huge New York landfill to recover methane gas presently being wasted. Using a technique developed in California, the utility expects to salvage 250,000 cubic feet of methane a day, enough gas to heat about 2,000 homes. Projected expansion would yield 2 million cubic feet of gas daily.

Government Programs

The Environmental Protection Agency, which sees sufficient energy in waste to light every home and building in the United States, is putting money into conversion plants. EPA is funding waste-conversion projects in six cities, ranging in size from a 150-ton per day wet-pulping materials recovery plant in Franklin, Ohio, to the 1,000-ton per day pyrolysis steam-generating plant in Baltimore. Other plants are in St. Louis, Missouri; San Diego, California; Wilmington, Delaware; and Lowell, Massachusetts. ERDA also has an aggressive bioconversion program going and sees in it the potential for appreciable savings in oil and natural gas.

Biofuels are generally nonpolluting, widely available, and at the moment largely wasted. To our credit, we are moving from just being producers of trash to users of energy from trash. Waste may someday

This plant converts garbage to steam to generate electricity. It will save about 17 million gallons of oil a year. *Federal Energy Administration*

provide more power for the United States than nuclear plants have managed in their quarter-century of development. It remains to be seen if we can successfully revert to a biofuel economy through concepts like the energy ranch, on land or afloat. For every expert who ridicules the idea, there seems to be one who is willing to try. At the very least, the potential is there—at a price, of course—should we come to the point of needing to cultivate Btu bushes not just for food but for fuel as well.

8. Solar Energy

It is common to say that before Copernicus the Earth was considered as the center of everything, but this belief doesn't hold up under careful inspection. From earliest times the sun was treated as a deity; the giver and sustainer of life. Such an appraisal is easily understood: the sun is still the source of most of our vision, all of our food and fiber, and practically all the energy we use. Without the sun there would be no warmth, no gravitational pull, no center to keep the system of planets together. So it should be no surprise that the sun represents—as it always has—a limitless source of useful, clean, and extremely safe energy.

It has been said that all of our energy is nuclear in origin and, with some exceptions, this is true by definition. Solar energy is the result of nuclear fusion taking place some 93 million miles away from us in the complete safety of outer space. So far away that even when occasional explosive flare-ups occur on the sun only our communications systems suffer.

Electromagnetic radiation, about which we lack complete understanding, travels through tens of millions of miles of vacuum, reaching us as a broad spectrum from very short X rays to very long radio waves. It is a vast outpouring of energy from a star consuming its substance at the prodigious rate of a million tons per second, and producing some 380 billion trillion kilowatts of energy. Earth of course intercepts only the tiniest portion of this outpouring, about one-thousandth of one-millionth, to be more accurate. Yet even this token amount comes to something like 85 trillion kilowatts. In that last statistic lies the reason for the sudden great interest in solar energy as an alternative source for a world embarrassingly near the bottom of the barrel of fossil and other fuels that once seemed abundant.

There are three basic kinds of solar energy: thermal, photovoltaic, and photochemical. The simplest to understand, and to use, is thermal

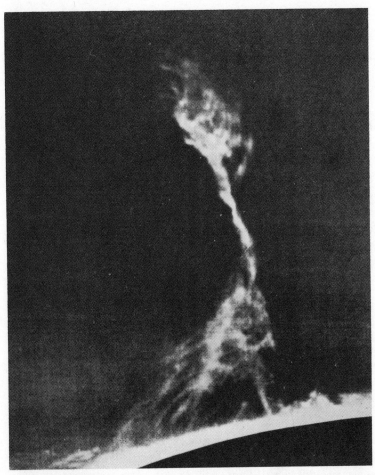

A "solar prominence" rises tens of thousands of miles above the sun's surface. *Walter Orr Roberts*

energy, the heat of the sun. Photovoltaic energy produces electricity from sunlight, obviously a more complex process. Even more complex is photochemical energy, which nature handles with ease but human beings only with great difficulty. We shall look at solar-energy applications in that order, beginning with the simplicity of solar thermal energy, the conversion of sunlight into heat.

Using the Sun's Heat

A paradoxical question that delights theorists concerns the riddle of whether there is sound from a waterfall if no one is around to hear it. Solar energy definitely doesn't amount to much until there is something to intercept it and convert it into one or another of the three forms of energy we have listed, in this case heat. The accepted term for such hardware is a "solar collector," although some engineers are belatedly trying to replace that with the more accurate "absorber" or "converter." Hold a piece of metal in the sun and it will be heated. Some of the electromagnetic radiation in sunlight is converted to sensible heat when intercepted.

Detractors of solar energy's potential often point out that sunlight is "diffuse," or thinly spread. The implication is that we shouldn't bother trying to harness anything so feeble. At first thought, this might seem to make sense, but a few statistics uncover the fallacy.

The fusion reaction going on in the sun's interior results in a power density of something like 65,000 horsepower per square yard at the sun's surface. Since this spreads out radially for 93 million miles, its density diminishes, of course. However, one square yard of Earth's surface, rooftop, or solar converter intercepting the direct rays of the sun receives about 1,000 watts or one kilowatt or about $1\frac{1}{3}$ horsepower. This is an appreciable amount of power in anybody's book. A roof 2,000 feet square with the sun directly overhead represents a potential of 222 horsepower! Of course, the sun is not always right overhead and sometimes it is not even visible. But discounting such limitations, and allowing for only about 10 percent conversion efficiency, there still remains enough solar power to handle all the needs of a modern house: heating, cooling, and hot water; plus electricity for appliances, lights, and so on.

One can visit Indian "condominiums" centuries old and smile at the primitive way our forebears lived. But these ancient cliff dwellings were designed far more efficiently than many residences today; they were solar heated in winter and protected from the sun in summer. Many present-day architectural triumphs are very poorly designed for air conditioning. In keeping out the summer sun we also bar the winter sun, which we would love to have. We burn increasingly expensive fuels to counteract the sun's heat in summer and duplicate it in winter.

Belatedly, some architects are considering the sun in the proper way, and then designing and installing solar converters and heat-transfer systems that will warm homes in winter. Surprisingly, the same equipment in some cases also refrigerates the home in summer. We will take up this bit of solar magic after we have mastered the more easily understandable heating process.

Sun tea is a homey backyard project, but I know people who swear that the beverage so made is far superior to that concocted with water heated on the stove. It has that "solar" flavor, is never strong, and also saves fuel in very tiny amounts. Here is one of the simplest solar water-heating applications.

If we put water in a five-gallon can painted black, the can—and the water—will get quite warm after standing in the sun for a length of time. The can eventually radiates away as much heat as it is receiving, however, and the water gets no hotter. In a properly designed solar collector, we circulate the water, or whatever it is that we are heating, and store it in an insulated tank.

To save a long story about the evolution of the "flat-plate collector," such devices today are generally made up of an absorber of some heat-absorbing material through which water is circulated; a frame to hold the absorber and insulate it; and a glass or plastic cover to let heat in but prevent it from escaping back through the glass. Add a storage tank and a method of circulating the heated water to the place it is to be used and we can "fuel" a variety of devices. The simplest is the domestic water heater, actually a rather old device used a century and more ago but largely forgotten with the coming of cheap fuel to heat the water more conveniently.

Swimming pools are a major application for solar heating. For years it has been customary to heat pools with natural gas and a few affluent owners do the job with electric power, more costly by three or four times. With the coming of the energy crunch, the solar pool heater came into its own. After all, it is a shame to invest thousands of dollars in a pool, and more hundreds a year on chemicals, cleaning, and filtering and only swim for a few months in the summer.

There are many types of solar pool heaters, from those quite like

Engineers check performance of flat-plate collectors using a new black chrome absorbing surface. *NASA*

the collectors used for space heating, to low-cost plastic panels specially designed for pool installation. The pool is almost an ideal application. Generally its circulating pump can carry water to and from the solar heater. The solar heater can also be in the form of a ramada, or sunshade, and thus do double duty. Water is a very handy medium for accepting solar heat and carrying it into the pool, and high temperatures are not needed. However, as one discouraged owner put it, heating one is something like trying to heat your house with the roof off! An uncovered pool can lose as many as 400,000 Btus in a single night, for example, far more than a house loses. Even during the day, while the sun is actually adding heat to the pool, much of that is lost in evaporation, convection, and radiation. A breeze makes the losses even greater. So the task is not just heating the pool, but keeping the heat in the water.

The answer is to install a solar heater large enough to replace heat constantly being lost. A rule of thumb says that the heater should have about half the area of the pool, although in severe climates it may need to be of equal area. Thus if a pool is 450 square feet in area, the solar heater should be a minimum of about 225 feet, say 10 feet by 22 feet. It should slope toward the south to better intercept the rays from the low winter sun. Another rule of thumb calls for a tilt angle equal to the local latitude plus 10 degrees. For a home at the 35th parallel of latitude, then, a solar collector should slope 45 degrees to the south. This is a pretty sharp pitch, and some pool owners prefer to compromise economics with esthetics and settle for a flatter slope.

Some flat-plate collectors suitable for house heating cost as much as $20 per square foot. At that price 220 square feet of a solar pool heater would bend the budget to the tune of $4,400, a cost about matching the pool itself. Fortunately, such a high-quality collector is not needed since water must be heated to only about 80 degrees F. So solar pool heaters can be more cheaply made and often are not glazed. In fact, an unglazed collector works better as a water *cooler* for hot summer conditions.

Some pool owners wondered why not make the pool itself the collector, and find a way to keep the heat in the water? The solution was almost ridiculously simple: a sheet of clear plastic of the cheapest kind, laid right on the water, does almost as much good as the average solar pool heater. Using both plastic and solar heater about doubles the temperature increase and thus prolongs the swimming season even more. The big problem is handling the awkward sheet of plastic; putting it on and taking it off between swims is inconvenient. Most people do not want to be bothered, so tricky automatic roll-up systems, and bubble plastic covers that roll up more easily, have been developed. Costing a fraction as much as a solar heater, they are becoming popular a bit more slowly than their more expensive counterparts, but someone is sure to make a breakthrough in convenience one of these days and revolutionize the pool-heating industry.

Hot water can also heat a house, as anybody who has adjusted a radiator must know. This task requires a larger area of solar collector, a larger storage tank, and a system of ducts and controls to do a proper job. But the basic principle is much the same as in the solar water heater, or even the Mason jar that brews delicious tea. Sunlight is converted to heat in the water, and the heat put to use.

Domestic hot water is an excellent solar-energy application since the required temperature is moderate, and thus permits a very efficient collector. At 140 degrees F., some commercial solar collectors operate at efficiencies as high as 80 percent.

A recent survey, done for the Internation Association of Sheet-metal Workers, by the way, reports that 83 percent of existing solar heating systems in the United States used water as the heat-storage medium. Another medium sometimes used is air. While water in solar collectors is often distilled or "buffered" in order not to corrode the plumbing, air requires no such treatment. Among the other advantages of air systems is the fact that leaks are not as messy as with a water system, and neither will air freeze. For these reasons, an increasing number of solar applications use air. After all, it is warm air that wafts over us to keep us comfortable in winter, and cool air in summer. Why heat water and *then* air?

Water has the advantage of being a good heat-storage medium, and a sizable tank in the basement or buried in the backyard can store enough solar energy to heat a house for several days. Storing equivalent warm air would take a large granary or perhaps a dirigible. But there are ways of solving the air-storage problem. One is to heat a bin of rock. Air is circulated through the solar collector on the roof and flows through a large bin of rock. Heat stored in this bin then supplies the house during the nighttime and other periods when the sun does not shine.

Most solar-heated buildings use more or less conventional flat-plate collectors to do the job. But there are alternatives well worth mentioning. One is the Thomason "trickling water" system. Instead of forcing water through the tubes of a closed system, Harry Thomason of Washington, D.C., pumps it to the top of his roof and lets it trickle down the corrugations of aluminum siding and flow by gravity into a buried storage tank. He has been doing this very effectively for nearly

The Thomason solar roof heats one of several homes designed and built by Dr. Harry Thomason of Washington, D.C. *Federal Energy Administration*

20 years in that city of severe winter weather, and providing most of the necessary heat.

Another solar engineer has a completely different but also effective water solar-heating system. Harold Hay is famed for his elegantly simple "solar roof," which is exactly that. Atop a sturdy corrugated steel ceiling, seven inches or so of ordinary water is contained in a clear plastic bag. Above the water are insulating panels of foam material. On sunny winter days, the panels roll back to uncover the water which soaks up countless Btus of energy. At night the panels cover the water, holding the heat in. The entire ceiling acts as a radiator to warm the house. No fans, pumps, noise, or fuel bills. And now for the miracle part: the Hay water roof also cools the house.

In summer the insulating panels remain over the water, keeping the unwanted solar energy out. At night, they roll back and the miracle of nocturnal radiation (loss of heat to the night sky) takes place. Hay came upon this remarkable phenomenon in his travels about Asia, where some natives, in their way wiser than modern scientists, made ice at air temperatures far above freezing. This they did by placing water in uncovered containers at night. Observant Americans surely noticed this phenomenon in our land too, but only Harold Hay seems to have done anything with it. Through nocturnal radiation the water may be cooled as much as 20 to 25 degrees F. below the surrounding air temperature. And the next day it transfers this coolness to (or rather absorbs heat from) the house below.

Hay's pilot model house at Atascadero, California, has been demonstrating its cost-free air conditioning for more than two years. Pacific Gas & Electric installed a meter for standby fuel, but that meter has yet to turn. The family of five put in the house by the Department of Housing and Urban Affairs rated the Hay system superior to conventional air conditioning. Best of all, this heating and cooling was free. As an added bonus, there should never be a roof fire on a water-roof house.

Harry Thomason used an idea similar to Hay's nocturnal radiation to help cool his house. Water trickling down his roof on summer

John Yellott and Harold Hay check performance of the latter's solar house design. Water roof both heats and cools the structure. *Solar Energy*

nights lost heat to the night sky and the next day helped cool the house. This same trick is used by owners of solar pool heaters who like their water refrigerated for summer comfort. Designers of air-circulating solar-heating systems have also found ways to cool houses with the same equipment. At night a fan blows cool night air into the rock bin, and by day the home owner draws on this stored "coolth" to keep him comfortable.

It is stretching the truth a bit to call such stratagems refrigeration, of course. But for the purist who insists on refrigerated air conditioning, there *are* ways of using solar heat to do this. One method is to run an engine with solar heat and operate a conventional compressor-type refrigeration system with it. A few such systems have been built but problems remain in this approach. More success has been attained with "absorption" refrigeration systems.

Absorption refrigeration is actually much older than the more common compressor refrigeration that most of us use in our homes. Absorption cooling is a classic example of serendipity, for in 1824, Michael Faraday, the English electrical genius, was doing an experiment aimed at producing liquid ammonia. He heated a container of silver chloride saturated with ammonia vapor. The ammonia vapor was driven off by the heat of a flame and condensed in another container. Satisfied, Faraday left his laboratory for a while and when he returned was surprised to find that as the liquid ammonia was reabsorbed into the silver chloride it left behind a residue of ice! Not only had he produced liquid ammonia, he had also invented absorption refrigeration quite by accident.

The first commercial refrigeration firm was named for Faraday. If you are acquainted with the Servel gas refrigerator, you have seen absorption refrigeration in action. Avoiding a complex scientific explanation, water heated to about 200 degrees F. serves to operate an absorption refrigeration system using ammonia and water, or lithium bromide and water. Hot-water energy takes the place of mechanical energy in the compressor. And the hotter it gets in summer, the more solar energy there is to operate the absorption refrigeration equipment to cool the house.

As this is written there are probably two dozen solar refrigerated residences and other structures in the United States. None provides

The Minneapolis-Honeywell solar van, heated and cooled by the folding collector mounted on roof. Under ERDA sponsorship, this test unit has toured the United States. *Minneapolis-Honeywell*

all of the necessary cooling, but with technical improvements coming along they may shortly do so. And they use the same solar collector that heats the building in winter.

The Solar Battery

Three-quarters of a century ago scientists were experimenting with remarkable passive devices that converted light into electricity. Interestingly, one of the first materials to be used was selenium, an element named for the moon goddess, Selene. Thus the first solar cells might have been called lunar cells instead. The silicon solar cell was pioneered by Bell Laboratories in the early 1950s. They were bona fide miracles, simply sitting in the sun and converting its rays directly into electricity that an appliance cannot tell from the utility kind, for an electric current is an electric current. A spin-off from the transistor effort, the first commercial solar batteries operated at a conversion efficiency of about 5 percent.

Solar cells have since had their efficiency increased to as high as 20 percent. But they still cost so much that the major user remains the federal government in its space effort. To buy enough cells to produce just one kilowatt, or about 1⅓ horsepower, today costs between $20,000 and $50,000. So the solar battery performs only a few specialized terrestrial jobs like powering remote radio or telephone stations and transistor radios.

The problem is that even though silicon is basically sand, and is the most plentiful element, solar cells require "single-crystal" silicon, an ultrapure type that is very costly to produce. Programs underway are hoped to cut costs by a factor of 10 and possibly another 10 after that. Basis for the hope to reduce solar-cell prices by a factor of 100 lies in a similar reduction made in transistor devices in a ten-year period after their development. This was done by government incentives and by a strong commercial market as well.

Along with the cost-cutting attempts, there is another approach that may help do the job. A magnifying glass concentrates a lot of sunlight onto a small area. If we focus this concentrated sunlight on a solar cell, we greatly increase the electricity coming out of that cell, unless it melts in the process, that is. Researchers have succeeded in cramming sunlight onto a solar cell at a concentration ratio of 1,725 to 1, or "1,725 suns" as the solar people term it. Of course, the cells must be cooled to keep them from melting down into liquid silicon or gallium arsenide or whatever they are made from. But researchers point out that such a concentration ratio on a square yard of solar cells would yield 250 kilowatts of electricity, about enough for the electric needs of about 250 city dwellers!

As with all wonderful ideas, there is a catch. To make the equivalent of 1,725 suns shine on a solar panel one must provide a lens or reflector with 1,725 times as much area as the solar cells. And a 1,725-square yard array would be a sizable and costly installation. However, it shouldn't cost as much as solar cells, so there may be a payoff there somewhere. When a combination of solar-cell price reduction and concentration techniques succeeds, we will have a new and limitless source of electric power for household and other needs. The solar cell used for the concentration of 1,725 suns to produce electricity was made of gallium arsenide, a more expensive type of cell. These cells cost about ten times as much as silicon, and the exotic materials required to make them are far more limited than silicon.

Varian scientist checks performance of solar-cell concentrator setup. Solar radiation is concentrated on cell more than 1,700-fold. *Varian Associates*

Among the other materials used for solar cells is cadmium sulfide. Such cells cost less because the material can be applied as a thin film. This is far cheaper than present silicon-cell techniques that consist of growing ingots, slicing wafers, and so on. Cadmium sulfide cells are less than half as efficient as silicon. But one could use twice as many cells and still do the job more cheaply if those cells cost less than half what silicon costs. A firm in Texas is studying a new mass-production technique of depositing cadmium sulfide on a thin glass surface, and predicts costs as low as six cents per watt! A kilowatt of power for only sixty dollars would usher in the solar electric with loud cheering from all sides.

Since cadmium is available in limited supply, and industry uses it for a variety of applications including plating, cadmium sulfide cells are not likely to figure largely in full-scale manufacturing programs. Neither are gallium arsenide or other exotic materials. It is most fortunate that silicon is an excellent material for solar cells, since silicon is the most plentiful element on Earth.

Solar cells are presently only about a $4 million-a-year industry, small in the scale of things. Thus government incentives are necessary to interest a company in the $50 million-a-year near-term production efforts being considered. ERDA has an aggressive Photovoltaic Electric Power Systems (PEPS) program. Although the goal seems to be very optimistic, considering that solar-cell prices have not been reduced appreciably in twenty-five years of research and development, costs are hoped to be cut to about $500 a kilowatt by 1980. Such a price would make them directly competitive with conventional power plants, some of which cost more than that now.

By the late 1970s a 200-kilowatt solar-cell power plant is to be tested, and by the early 1980s a 400-kilowatt plant will follow. As soon as possible after that a 600-kilowatt plant will be built. These are very small compared with the 10-megawatt (10,000-kilowatt) solar-thermal power plants. But this does not minimize their usefulness. Indeed, there is a decided advantage to being able to produce solar power plants economically in small sizes, since they could then qualify for rooftop installation on residential, business, and industrial buildings. A solar-thermal power plant must be built on a much larger scale to be economically attractive.

Estimates suggest that PEPS installations with an efficiency of 10 percent could supply 75 percent of the demand for electric power in the year 2000, using only 1 percent of the land areas of the continental United States. For comparison, farming now takes up about 15 percent. PEPS has the added advantage that solar-cell arrays could be installed on existing structures rather than on real estate specially set aside for them. The government's Project Independence blueprint suggests that such solar-cell power arrays would make no significant adverse impact on the environment.

Following the production of $500-per-kilowatt solar-cell arrays, according to the timetable, would come others costing only $100 per kilowatt. Such an achievement would practically guarantee the rapid development of large-scale installations. Indeed, the federal goal is production of 50,000 megawatts annually by the year 2000. This would match present hydroelectric power output and provide about 8 percent of electrical needs.

Photochemical Reactions

The third major category of solar energy is photochemical. We have been using photochemistry on a small scale in photography for many decades, thus the sun is already saving us money in flashbulbs and other artificial lighting, if we look at it this way. However, there is far more potential in photochemistry than that modest contribution.

While few can explain photosynthesis, most of us are to some extent aware that this wonderful process fuels growing plants to produce carbohydrates. These carbon dioxide and water compounds are food or fuel, whichever way we want to consider them. In an earlier chapter we looked at a specialized form of solar energy use called bioconversion. For now, however, we restrict our discussion to *inorganic* processes in converting solar energy to fuel, or to power.

There is a long-known photochemical process called the Hill reaction, in which a weak electric current can be produced by the action of sunlight on a liquid. Experiments show that it should also be possible to produce fuel from water by irradiating it with light. Water, of course, is hydrogen and oxygen. If we could separate these gases by the photolytic process (we can easily do it now with electricity, or electrolysis), we would have two very useful fuels. Many energy authorities believe that we shall one day have a "hydrogen economy" rather than a fossil-fuel or nuclear economy. But hydrogen must be produced, and that requires an energy source.

The economical photolysis of water would be a brilliant success for solar energy utilization. The excellent gaseous fuels would give us a method of storing solar energy easily, and storage is one of the biggest problems with this promising source. It is available for only part of the day, and then we must use batteries, pumped water storage, or the like for continuous power.

Hydrogen and oxygen can be burned or otherwise converted into power. They can also be combined in a fuel cell to produce electricity very efficiently, particularly when compared to "heat engines," which seldom deliver more than about one-third of the energy in their fuel. The fuel cell operates at efficiencies that theoretically approach 100 percent. And the only byproduct of this conversion of gases to electricity is pure water, which can then be broken down again by solar energy to begin the power cycle anew.

Of the three basic solar energy methods, photochemistry remains the least understood, and the most difficult challenge for scientist and engineer. However, the potential payoff is correspondingly large, and researchers are working diligently in this specialized branch of solar energy.

The Solar Houses Are Coming

In 1974 Representative Mike McCormack, Democrat of Washington, introduced a bill calling for a solar heating and cooling demonstration across the United States. McCormack's concept was for the federal government to underwrite the solar portion of 2,000 solar-heated buildings and 2,000 solar-heated and cooled buildings. The Solar Heating and Cooling Demonstration Act passed late in 1974, and ERDA, aided by NASA and HUD, is now in the process of carrying out the program.

With or without the Demonstration Act, which is funded in the amount of $60 million, solar heating and cooling seems about to make it anyway. Long before ERDA was created, the National Science Foundation funded four sizable solar-heated school projects, in Minnesota, Massachusetts, Maryland, and Virginia. While each cost far more than it would be economically feasible to spend on a continuing scale, the projects must be considered as research and development. Across the country, many architects are now looking more and more seriously at solar air conditioning as a way to save not only scarce fuels, but money as well.

Solar scientist William Shurcliff periodically publishes a survey of existing solar homes and buildings. His twelfth edition, printed in 1976, lists more than 200 such structures. However, solar heating and cooling is coming on so fast now that it is impossible to keep up with the numbers. For example, Arizona has at least fifty solar homes and other buildings completed or under construction, including the nation's first solar car wash and first solar-heated astronomical observatory.

So bullish are Colorado solar energy developers that one declared that the total solar market by 1985 could surpass the present total market for the automotive industries. An official of Solaron, the most active solar design and installation firm in the nation, announced at the National Building Show in Chicago late in 1975 that the 1,000 or so solar buildings in 1975 would rise to as many as 6,000 in 1976 and that by 1980 there could be as many as 500,000, with a value of installed solar equipment of $2.5 billion. Ultimately, he pointed out, there should be about 40 million candidate buildings for solar installations.

Frank Zarb, head of the Federal Energy Administration, announced that solar equipment produced by manufacturers in 1975 when installed would result in savings of about 600 barrels of oil daily, or 219,000 barrels a year. This is still just a drop in the bucket but Zarb also pointed out that his agency is working to stimulate a solar market that will save one million barrels a day.

ERDA's publication 75A, released late in 1975, is surprising evidence of how fast a solar industry is forming. This document, listing manufacturers of solar hardware, runs to 400 pages and details hundreds of suppliers in the United States. Some are large companies like General Electric, Pittsburgh Plate Glass, Reynolds Aluminum, Revere Copper and Brass, and Grumman Aircraft. Others are newcomers, many of them tiny operations just getting off the ground. Inevitably there will be solar equipment sold and installed that fails to do the job it is claimed to do. This will be because of ignorance on the part of buyer and seller, and in some cases because of opportunists sensing solar energy as another aluminum siding pot of gold.

Concern for this problem is one reason for ERDA's agonizingly circumspect approach. Only a competent, qualified solar expert can put together the kind of solar heating or cooling proposal that will be funded. The federal government also has established a comprehensive set of "interim criteria" for solar hardware. Put together by the National Bureau of Standards with much help from solar authorities around the country, these criteria are a yardstick against which standardized testing can be measured.

The most elegant solar-heated and solar-cooled residence in the world. Even the pool is warmed and cooled as needed. *Copper Development Association*

One of the first solar-heated and solar-cooled schools in the United States. AAI Corporation installed the solar collectors on this Timonium, Maryland, elementary school. *Aircrafts Armaments, Inc.*

Bridgers and Paxton, who built their first solar-heated office building in 1956 at Albuquerque, recently completed the largest solar-
heated and cooled building in the United States, a new agriculture
building for the Southern New Mexico University at Las Cruces. To
select the solar collectors, Bridgers and Paxton canvassed the industry and finally chose a unit manufactured in Israel. This collector,
with a track record of fifteen years of operation, supplies most of the
heat for the new school building. But a number of newer, U.S.-built
and more sophisticated, tracking and concentrating collectors were
also installed to provide the higher temperatures needed for the refrigeration system. Bridgers and Paxton are also heating and cooling
a large community college near Denver, Colorado, with solar energy.
The new college (with 50,000 square feet of solar collectors) will be
the largest solar building in the world.

U.S. dominance in solar development may not last long, inciden-
tally. Saudi Arabia's interest was demonstrated in the 1975 visit to
the United States of Prince Turki al Faisal of the royal family on a
mission to enlist help for solar development in his country. Senator
Paul Fannin (R., Arizona), a key energy legislator on the Interior
Committee, reported that the prince mentioned having $60 billion to
spend and complained of slowness on the part of U.S. federal agencies in solar progress. While on his visit, the prince delivered the first
$150,000 of a $625,000 grant that his country is providing the
Reston, Virginia, Terraset Elementary School for a solar air-condi-
tioning system. All that the Saudis asked in return was permission to
study the building as a model for solar cooling which could be used
in their country.

Solar Power Plants

Thus far we have discussed solar energy applied to on-site uses in
heating water and heating and cooling homes and buildings. The low-
technology hardware that suffices for these applications cannot also
produce electricity, and our life style is tied tightly to this form of
power. While many dream of windmills on every roof, or of solar
batteries at one dollar a square foot shingling the homes in subdivi-
sions, neither of these approaches is likely to succeed anytime soon.

Before spending too much time bemoaning that news, consider the solar-electric power plant: not just a few panels of solar cells making dribbles of electricity but large-scale facilities tied in with the power grid and making megawatts out of sunshine.

In Arizona all three major utilities are involved in serious solar energy power-generation projects. And the president of the largest utility is also chairman of Arizona's Solar Energy Research Commission. Despite the cries from young people, and some old folks too, that the utilities should not be allowed to rip us off on "free" solar energy, these companies are the logical candidates to build the new generation of solar power plants. If they are to invest billions of dollars (for solar energy hardware is not free), they are entitled to a profit on that investment.

We have devoted a quarter of a century and sizable sums of money in the research and development of such "fusion" power plants, although success has not yet rewarded this time and money expended. It is remarkable that those with the imagination and ingenuity to duplicate the powerhouse in the sun have been so tardy in learning to use the energy the sun itself produces.

John Ericsson, the brilliant Swede who built the *Monitor* of Civil War fame and many other remarkable things as well, was one of the few who dreamed of tapping the energy pouring from the "huge powerhouse in the sky." Ericsson actually built sizable solar engines developing a few horsepower, but the world was not ready for this application.

A handful of others followed his lead. The French scientist August Mouchot progressed from solar cookers to a small solar engine which powered a small printing press among its demonstrations. Inventors in the United States tried very hard to interest others in solar engines in sizes ranging from 4½ horsepower and up. Frank Shuman actually built a huge solar steam engine that produced something like 60 horsepower to pump irrigation water at Meadi, on the outskirts of Cairo, Egypt. The year was 1913, but World War I plus apathy doomed the marvelous solar engine.

Dr. Charles G. Abbot, a former Secretary of the Smithsonian Institution and the father of solar-energy research and development in the United States, built a small solar-powered steam engine in the 1930s and demonstrated it by providing the electricity for a coast-to-coast radio broadcast. A couple of decades went by before some Italian engineers rediscovered the solar irrigation pump, capable of converting sunlight into horsepower for light pumping tasks. But it was a farsighted and diligent French scientist who set the stage for the present revolution taking place in solar power generation.

Dr. Felix Trombe possessed not only a knowledge of the sun's potential but also the ability to coax sufficient sums from the French government in post-World War II years to build a sizable solar furnace in southern France. This device was 35 feet in diameter and produced about 100 kilowatts of thermal energy at temperatures of several thousand degrees Fahrenheit. In the late 1960s Trombe followed up this solar success with a monster furnace at Odeillo, some 115 feet by 165 feet, and served by a large array of mirrors on a nearby hillside. This furnace produced an impressive one megawatt, or 1,000 kilowatts, of thermal energy at temperatures as high as 5,600 degrees F.

A megawatt of heat energy will produce about 300 kilowatts of electricity, and so for the first time solar science had produced a device capable of making enough electricity for a small village. Trombe has not yet put his furnace to that use, but he is still smelting metals and doing other tasks with solar heat.

Russian scientists first proposed what today is known as the "central tower" solar power plant. Back in the 1940s they made drawings of a high tower, mounting a steam boiler, surrounded by several concentric railroad tracks. On the tracks railroad flatcars mounted large mirrors, all aimed at the tower. The idea was simple: just bounce the sun's rays off the mirrors and reflect them up into the tower's boiler cavity. The resulting steam would drive a conventional engine that in turn could run a generator and make electricity.

The response from most of the scientific community ranged from raised eyebrows to outright laughter. In the rare event that it might work, who needed solar energy anyhow? Fossil fuels were cheap and the cost for a million Btus was about twenty-five cents. To capture that much solar heat would require a great engineering effort and quite a bit of cash. If the Russians ever built their proposed central tower solar power plant, they kept the secret well, and the central tower idea faded away and was not heard of again for several decades.

Frank Shuman, who had spent his life and his fortune trying to convert sunlight into power, in 1909 proposed a solar energy farm covering several acres. He thought he could flood a shallow basin with water, cover the pond with paraffin to keep in the heat, and warm the water enough to run a low-efficiency steam engine. His scheme ran decades ahead of the technology available to him. But it was the same idea that made headlines in Arizona in 1970 when Dr. Aden Meinel and his wife Marjorie Meinel of the University of Arizona announced their "desert solar power farm." Dreamed on a scale that would have delighted Frank Shuman, the Meinel design covered not acres, but 5,000 square miles. And 5,000 miles of Arizona sunshine would be converted into a billion kilowatts, enough to take care of the entire nation's needs in the year 2000, which was the estimated completion date for the power farm.

With Btus still selling for bargain prices, and the Arab oil boycott still several years off, energy experts paid about as much attention as they did to others who warned that we were nearing the end of the cheap energy kick. The best the Meinels could do was to get enough funds from the National Science Foundation to do an exhaustive study of the proposed solar farm and to build a few pieces of test hardware to check out their ideas.

The Meinels' plan was to use flat-plate solar collectors to raise the temperature of water in a series of pipes high enough to operate a steam turbine. Meinel has found it impossible as yet to achieve the very high temperatures between 800 and 1,000 degrees he had originally hoped for. When he developed special thin films that did produce the high temperatures in his heat pipes, the films themselves couldn't take the fierce heat and degraded quickly. So Meinel began to look at concentrating collectors. One present approach is very much like the Shuman engine in Egypt, and a solar irrigation pump proposed by Meinel to develop about 200 horsepower would use parabolic concentrators of just about the size Shuman had built. Here was a good indication of technological advance in the more than half a century since Shuman achieved his dream; Meinel would realize three times as much power from solar collectors of the same size.

Others began to get into the solar power act, as the National Science Foundation began making important money available for such research. One firm was Minneapolis-Honeywell of Minneapolis, Minnesota, surely evidence of large amounts of faith considering that state's small share of sunshine! Senator Hubert Humphrey, who had long been an advocate of solar research and who introduced a bill to set up a solar-energy research operation in 1962, doubtless had something to do with this activity in his home state. The Honeywell design was a linear concentrating collector, and exhaustive testing was begun at Arizona's Desert Sunshine Exposure Tests, Inc.

A number of solar power designs were suggested, and some testing was done. Then quite suddenly the old Russian central tower concept became the darling of the Atomic Energy Commission. Resurrected by scientists at Houston University, the idea differed mainly in scope and sophistication. A tower some 500 meters high was proposed. And instead of mirrors mounted on railroad cars, the Houston group suggested pedestal-mounted mirrors tracking the sun automatically.

Almost overnight it was decided by those with federal money to spend that the concepts of Meinel and others that depended on a distributed system of heat collection would not succeed because so much heat would be lost in moving the heated fluid from the farthest reaches of the solar farm to the boilers. The central tower, on the other hand, would not convert sunlight to heat until the sun's rays reached the central boiler.

The central tower is moving along on a three-pronged program that will cost at least $100 million before a prototype 10-megawatt plant is operating. Completion is planned for 1980. In 1976, the engineering firm of Black and Veatch is building a 5-megawatt thermal-test central tower at Sandia Laboratories in Albuquerque, New Mexico.

Like a hydropower plant, a solar plant must be built where there is energy. It would therefore seem good business to build the pilot 10-megawatt solar plant not at the North Pole or in a Newfoundland fog bank but in an area of high solar radiation. So there is great jockeying going on between New Mexico, Texas, California, and Arizona. A 10-megawatt plant would handle the electric needs of a town of between 5,000 and 10,000 people, so it is not a solar toy. ERDA talks

This test facility for the solar-electric power plant will be completed in late 1977. *Black & Veatch*

hopefully of completing a 50-megawatt solar power plant soon after 1980.

In essence, the central tower is a large boiler, in which pressurized steam will be generated at around 1,000 degrees F. Surrounding the tower are acres of large mirrors, automatically controlled by computers and solar-cell tracking devices to keep the sun's rays focused on the boiler. According to experts, automated control of even 5,000 mirrors used in some designs will pose no problem, either technically or economically, because of the low cost of simple microcomputers and processors.

ERDA is not the only agency working toward the solar electric plant. The Electric Power Research Institute (EPRI), a utility-funded organization, has a sizable solar-research program and is funding a Boeing Aircraft Company central-tower project. For ERDA, Boeing is designing only mirrors. Its approach was unique among the others, with each mirror covered by a protective plastic bubble. Concurrent with the 10-megawatt design work is work on a 100-megawatt plant, sizable enough to qualify as an honest-to-goodness commercial electric power plant.

The Total Solar Community

Flexibility is among solar energy's more important advantages. It works in the backyard, and if the solar electric power program succeeds, it will work in huge utility grids. There is a very interesting application in between these two extremes, called the "total solar energy system." Among those working on this intermediate concept are researchers at Sandia Laboratories.

The basic idea of the total energy system is not to use the sun to provide *either* electric power *or* hot-water heating, but to do both chores. In generating electric power, there is a great waste of energy in "low-grade" heat. This heat is at a temperature that could provide domestic and industrial hot water, as well as space heating and cooling. For a commercial power plant to transport such waste heat to a city would not be practical, however, and this is where the moderate-size community total-energy concept is attractive.

The Sandia researchers plan a complex of 1,000 homes, clustered

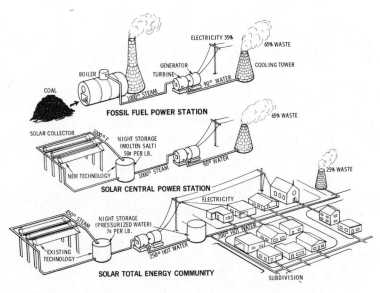

ELECTRICITY 35%
65% WASTE
GENERATOR
COOLING TOWER
TURBINE
BOILER
90° WATER
COAL
1000° STEAM
FOSSIL FUEL POWER STATION
65% WASTE

SOLAR COLLECTOR   1000°F
NIGHT STORAGE
(MOLTEN SALT)
50¢ PER LB.
90° WATER
25% WASTE
NEW TECHNOLOGY   1000° STEAM
SOLAR CENTRAL POWER STATION
ELECTRICITY

450° STEAM
NIGHT STORAGE
(PRESSURIZED WATER)
7¢ PER LB.
200° HOT WATER
EXISTING
TECHNOLOGY   250° HOT WATER
SOLAR TOTAL ENERGY COMMUNITY   SUBDIVISION

How the "total energy" concept will make solar power facilities more efficient. *Sandia Laboratories*

around a community center. A central solar power supply would generate electricity for lights and appliances, plus hot and cold water that would be circulated to the individual homes. Thus, rather than using only about 30 percent of the available solar energy and wasting the rest, the total energy system would recover perhaps 60 percent or more. Here would be the best of both worlds, unattainable with either the house-top application or the very large, central power plant. Should the solar total-energy program succeed, within ten years a total of 750 1,000-home communities might be completed, saving an estimated 34 million barrels of oil a year.

Of all the alternative energy sources, direct solar energy is obviously the one with the most promise. It alone can eventually provide *all* our needs for as large a population as it is pleasant to consider, and on an inexhaustible basis. While it is not to our credit that we did not develop solar energy rather than other energy economics until we had no other choice, it is comforting to know that it is there waiting for us to put to work. The future of solar-energy applications is as bright and as long-lasting as the sun itself.

9. Crossroads for Energy

The Energy Research and Development Administration
(ERDA) was established under the Energy Reorgani-
zation Act of 1974 to bring together federal activities
in energy research and development and assure coor-
dinated and effective development of all energy sources.
Congress, in authorizing ERDA for this purpose, stated
as a national goal: ". . . effective action to develop,
and increase the efficiency and reliability of use of all
energy sources to meet the needs of present and future
generations, to increase the productivity of the national
economy and strengthen its position in regard to international trade, to make the Nation self-sufficient in
energy, to advance the goals of restoring, protecting,
and enhancing environmental quality, and to assure
public health and safety." ERDA officially began this
task on January 19, 1975 under Executive Order 11834.

ERDA, November 21, 1975

Understandably, this is an age for doom-crying. One of the ironies of
our wasteful consumption of material and energy is paralleled by a
great popularity of books assuring us that the sky is not only falling
but about to reach our deserving heads. Given our dismal record,
such scenarios are very easy to write, and about as easy to believe.
Any inventive sophomore can quickly draw curves that show us running out of raw materials, fuel, space, and just about everything but
pollution and sin. Thus it might be popular to end this book on a
down note, predicting a horrible future for the world and particularly
for the developed countries. (At least the others will not have as far
to fall.)

For the first time in the history of our nation, population is dropping—
and precipitously. We were warned of what would happen and, sure

enough, tragedy has taken its toll in a society that could not believe the warning. Slowly we are becoming once again an agrarian economy, tending raggedy crops with muscle power, hoarding anything burnable from the fields to use in stoves—and occasionally to run a primitive steam engine. A few of the more clever coax methane gas out of home-built digesters, and use this fuel to cook with, and occasionally to drive the rickety old car to town and back.

We have burned up the last tanks of gas and the last barrels of oil, and the nuclear plants are cooling to the temperature of the air around them as the precious uranium has all been converted to heat and radioactive waste. Long ago the last *Concorde* and even the economical 747's were mothballed, or cannibalized for metal and other materials. No trains run, and all highways are lined with abandoned automobiles, trucks, and buses, many of them gutted in maddened fury by those who had to give them up.

In farm fields tractors sit half-buried in mud or dust. There is still waterpower, of course, but long ago competing groups dynamited transmission lines to the ground in their battles for what little electricity remained. Of course there is a black market in gasoline and oil, but only the better-off can afford such luxuries.

In an earlier energy revolution waterpower supplemented the power of human and animal muscles.

It is a satisfying picture for some. Out we go, not with the bang that antinuclear people had prophesied, but with a starved whimper. Cold, ill, and hungry in our rags and the growing darkness, just as we were warned. But that perhaps deserved fate is not likely to overtake us. Our luck continues to hold.

The current energy crisis is but the latest in a number of roughly similar crossroads in civilization's journey through time and the world. The first such crisis came far before recorded history, centering on food. The meat eaters became greedy and decimated the once plentiful supply of game. But instead of withering on the vine, the human race developed agriculture with revolutionary results.

Another crossroad in the human struggle was recorded more clearly for us in history, as an energy crisis reared its head. Yet when Rome was faced with a shortage of slave power the city fathers did not revert to cavemen. Instead they belatedly exploited waterpower. The marvelous system of aqueducts was tapped for power in addition to water, and mills and other machines were turned by nature instead of by slaves. Windmills later added to the power supply and humanity became addicted to this effortless kind of living. Such machines in time led to wood-fired steam engines, and before long there was another energy crisis when the wood was all used up. Despite the doom-cryers, who were vocal in those days too, civilization did not go to hell. Instead its members got busy digging coal to start another revolution. Later, oil was added, and so was gas.

Today, in spite of coal, oil, gas, and a newcomer called nuclear energy, we are in an energy bind again. But because we are stubborn types, this latest crisis too will likely be resolved. There is one critical difference this time, however. In ages past, the shift from one energy source to another was not accomplished quickly. Instead, it stretched out over a very long period of time. Agriculture, for example, did not transform the whole world at once and some areas lagged centuries behind the pioneers. Even the most recent change from one energy source to another has never taken place in less than about sixty years. This time, however, because we have peaked so fast in our use of fossil fuels, we don't have sixty years to switch. We have only a decade or two to get the job done, as a matter of fact. But it seems likely that human beings who can put men on the moon, produce color

After providing fuel for the heat engine that revolutionized the power field, wood in turn gave way to coal and another revolution.

Ultimately the sun may provide all the energy we need—for as long as we need it. *Robert McCall*

TV, and sell electronic calculators for $8.88 can bring off another miracle.

As is obvious from the preceding chapters, we have a number of things going strongly for us: the Earth, the wind, the sun, and the water. There are other helpful factors as well. The Energy Research and Development Administration is one of them. In times past about the only energy policy this country seems to have had was not to have a policy. The oil and gas interests did their thing, as did the older coal industry. When nuclear energy came on the scene it created the most complex energy sector of all time. Now, helpfully, all energy sources are under the ERDA umbrella. True, most ERDA people recently wore AEC hats, but they are learning that there are other ways to go. And new people are coming into the ranks to help them with those new ways. Solar people and geothermal people. Windmill builders and ocean-energy proponents.

For the first time, the United States has the beginnings of an energy policy. One early step was called Project Independence, with even a blueprint for that independence from imported fuels. The main thing about the ERDA approach is that it does not put all our eggs in the same old energy basket. Even the most dedicated nuclear physicist realizes now that nuclear power most likely will never serve the bulk of our needs—as well-meaning pioneers once promised. Coal, while it could serve for some time, will not do the job alone either. Nor will oil and gas. But there is growing help from the new—and ever so old—alternatives like geothermal energy, and wind power, and the tides, and the heat of the ocean, and the direct heat of the sun. The sun alone has the potential, should we ever need it, to provide practically all our power needs for as far in the future as one could care to peer. It's our move, and the prospects are encouraging.

Further Reading

Geothermal

Energy for Survival. Wilson Clark. Garden City, N.Y.: Doubleday & Company, 1974.
Energy in the Future. Palmer Coslett Putnam. New York: Van Nostrand, 1953.
Energy Primer. Portola Institute, 1974. 558 Santa Cruz Avenue, Menlo Park, CA 94025.
Geothermal Energy. William W. Eaton, Energy Research and Development Administration. Washington, D.C.: U.S. Government Printing Office, 1975.
The Mother Earth News Handbook of Homemade Power. New York: Bantam Books, Inc., 1974.
The Potential of Low Temperature Geothermal Resources in California. California Division of Oil and Gas, 1416 Ninth St., Sacramento, CA 95814, $2.50.
Power Production. Hans Thirring. London: Harrap, 1956.

Hydropower

Electric Power in America. Robert McCaig. New York: G. P. Putnam's Sons, 1970.
Hydroelectric Power Resources of the United States. Federal Power Commission. Washington, D.C.: U.S. Government Printing Office, 1972.
Producing Your Own Power. Carol Hupping Stoner. Emmaus, Pa.: Rodale Press, 1974.
Windmills and Watermills. John Reynolds. New York: Praeger, 1970.

Tidal

Tidal Power. T. J. Gray and O. K. Gashus. New York: Plenum Press, 1972.

Sea Thermal Energy

Solar Sea Thermal Energy. Committee on Science and Astronautics, U.S. House of Representatives. Washington, D.C.: U.S. Government Printing Office, 1974.

Wind

Electric Power from the Wind. Henry Clews. Solar Wind Co., Box 7, East Holding, ME, 1973.

Power from the Wind. Palmer Coslett Putnam. New York: Van Nostrand Reinhold Co., 1948.

Wind Power. Charles A. Syverson and John G. Symons. Box 233, Mankato, MN, 1973.

Bioconversion

Bio-Gas Plants: Designs with Specifications. Ran Box Singh. Gobar Gas Research Station, Ajitmal, Etawah (U.P.) India, 1973.

Practical Building of Methane Power Plants. L. John Fry. Whole Earth Truck Store, 1974. 558 Santa Cruz Avenue, Menlo Park, CA 94025.

Solar

The Coming Age of Solar Energy. D. S. Halacy, Jr. New York: Harper & Row, 1973.

Direct Use of the Sun's Energy, Farrington Daniels. New Haven: Yale University Press, 1964.

Solar Homes and Sun Heating. George Daniels. New York: Harper & Row, 1975.

Appendix: Table of Metric Equivalents

U.S. to Metric	Metric to U.S.

Linear Measure

U.S. to Metric	Metric to U.S.
1 inch = 25.4 millimeters (exactly)	1 millimeter = 0.0393701 inch
1 inch = 2.54 centimeters (exactly)	1 centimeter = 0.393701 inch
1 foot = 0.3048 meter (exactly)	1 meter = 3.28084 feet
1 yard = 0.9144 meter (exactly)	1 meter = 1.09361 yards
1 mile = 1.609344 kilometers (exactly)	1 kilometer = 0.621371 mile

Square Measure

U.S. to Metric	Metric to U.S.
1 square inch = 6.4516 square centimeters (exactly)	1 square centimeter = 0.1550003 square inch
1 square foot = 0.092903 square meter	1 square meter = 10.7639 square feet
1 square yard = 0.836127 square meter	1 square meter = 1.19599 square yards
1 acre = 0.404686 hectare	1 hectare = 2.47105 acres
1 square mile = 2.58999 square kilometers	1 square kilometer = 0.386102 square mile

Cubic Measure

U.S. to Metric	Metric to U.S.
1 cubic inch = 16.387064 cubic centimeters (exactly)	1 cubic centimeter = 0.0610237 cubic inch
1 cubic foot = 0.0283168 cubic meter	1 cubic meter = 35.314725 cubic feet
1 cubic yard = 0.764555 cubic meter	1 cubic meter = 1.30795 cubic yards

Index

Page numbers in *italics* denote illustrations.